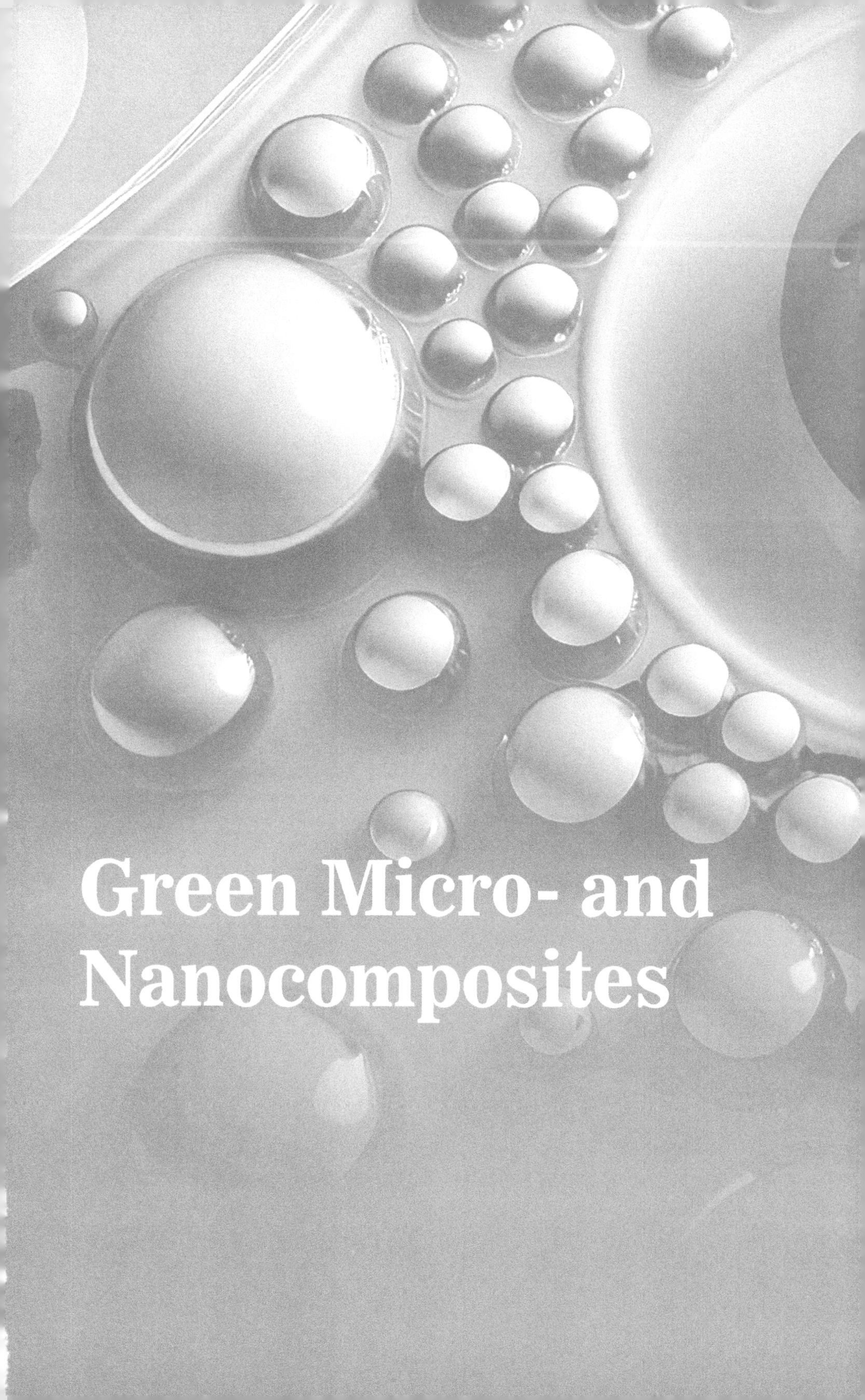

Green Micro- and Nanocomposites

Green Micro- and Nanocomposites

edited by

Sabu Thomas
Abitha V. K.
Hanna J. Maria

JENNY STANFORD
PUBLISHING

Published by

Jenny Stanford Publishing Pte. Ltd.
101 Thomson Road
#06-01, United Square
Singapore 307591

Email: editorial@jennystanford.com
Web: www.jennystanford.com

British Library Cataloguing-in-Publication Data
A catalogue record for this book is available from the British Library.

Green Micro- and Nanocomposites

ISBN 978-981-4968-79-9 (Hardcover)
ISBN 978-1-003-42756-8 (eBook)

Contents

Preface

In recent years, there has been a growing global concern about the environmental impact of materials and on the urgent need for sustainable solutions. In response to this challenge, the field of green materials has emerged as a promising avenue for developing eco-friendly alternatives that minimize the ecological footprint while maintaining or even enhancing performance. One such area of research is the development of green micro- and nanocomposites, which combines the principles of green chemistry and advanced materials science to create sustainable and functional materials.

This book provides a comprehensive overview of the state-of-the-art advancements in the field of environmentally friendly micro- and nanocomposite materials. It brings together the collective knowledge and expertise of researchers and practitioners from various disciplines, including materials science, chemistry, engineering, and sustainability.

The objective of this book is to provide a comprehensive overview of green micro- and nanocomposites, offering a holistic understanding of their synthesis, properties, and applications. It encompasses a wide range of aspects, including green synthesis methods, characterization techniques, mechanical and thermal properties, electrical conductivity, and barrier performance of micro- and nanocomposites.

Chapter 1, Green Micro- and Nanocomposites, serves as an introduction to the subject, discussing the importance of green composites and their potential to revolutionize various industries. It sets the stage for the subsequent chapters, which delve deeper into specific aspects of green composites.

Chapter 2, Biodegradability of Green Composites: Mechanisms and Evaluation Methods, explores the fascinating world of biodegradability, examining the mechanisms involved and the methods used to evaluate the biodegradation potential of these materials. Understanding the biodegradability of green composites is crucial for their sustainable use in various applications.

Chapter 3, Green Composites Reinforced with Cellulose, focuses on the utilization of cellulose, a renewable and abundant resource, as a reinforcement in green composites. The chapter discusses the preparation, properties, and applications of cellulose-based composites, highlighting their unique advantages and challenges.

Chapter 4, Green Composites Reinforced with Chitin and Chitosan, explores the potential of chitin and chitosan, natural polymers derived from shellfish, as reinforcements in green composites. The chapter investigates the properties, processing techniques, and applications of these composites, shedding light on their immense potential in various industries.

Chapter 5, Green Composites based on Polyhydroxyalkanoates, focuses on the applications of polyhydroxyalkanoates (PHAs), biodegradable and biocompatible polyesters produced by microorganisms, in green composites. The chapter delves into the synthesis, properties, and applications of PHA-based composites, providing insights into their wide-ranging uses.

Chapter 6, Green Composites Based on Poly(Lactic Acid), explores the utilization of poly(lactic acid) (PLA), a renewable and biodegradable polymer derived from plant sources, in green composites. The chapter discusses the processing techniques, properties, and applications of PLA-based composites, highlighting their potential to replace conventional petroleum-based materials.

Chapter 7, Green Composites Based on Protein Materials, examines the use of protein materials, such as soy protein, zein, and gluten, as reinforcements in green composites. The chapter explores the preparation methods, properties, and applications of protein-based composites, emphasizing their biocompatibility and potential applications in the biomedical field.

Chapter 8, Surface Modification of Biobased Polysaccharide Nanoparticles via Grafting, focuses on the surface modification techniques employed to enhance the properties and performance of biobased polysaccharide nanoparticles. The chapter discusses various grafting methods and their impact on the composite materials, paving the way for tailored properties and applications.

This book will be a valuable resource for researchers, scientists, engineers, and professionals interested in the development and applications of green micro- and nanocomposites. By providing a comprehensive understanding of the synthesis, properties, and applications of these materials, it aims to inspire further advancements in the field and contribute to a sustainable future.

We acknowledge the collaborative effort of the authors, editors, and reviewers who have contributed to this book. Their expertise and dedication have ensured the quality and relevance of the content of the book.

Sabu Thomas
Abitha V. K.
Hanna J. Maria
Summer 2023

Chapter 1

Green Micro- and Nanocomposite Materials

Abitha V. K.,[a] Ajay Vasudeo Rane,[b] Deepti Yadav,[c] Hanna J. Maria,[d] and Sabu Thomas[a,d,e]
[a]*School of Chemical Sciences, Mahatma Gandhi University, Kottayam-686560, Kerala, India*
[b]*Composite Research Group, Department of Mechanical Engineering, Durban University of Technology, Durban-4000, South Africa*
[c]*Department of Biotechnology and Food Science, Durban University of Technology, Durban-4000, South Africa*
[d]*School of Energy Materials, Mahatma Gandhi University, Kottayam-686560, Kerala, India*
[e]*International and Inter University Centre for Nanoscience and Nanotechnology, Mahatma Gandhi University, Kottayam-686560, Kerala, India*
abithavk@gmail.com

Environmental issues throughout the world such as rising sea levels, rising average global temperatures, melting polar ice caps, and fast depleting petroleum resources have increased the burden on individuals and industries to find alternative resources to fulfil their

Green Micro- and Nanocomposites
Edited by Sabu Thomas, Abitha V. K., and Hanna J. Maria

ISBN 978-981-4968-79-9 (Hardcover), 978-1-003-42756-8 (eBook)
www.jennystanford.com

demands (Fig. 1.1). Key global challenges include how to respect the environment and enhance living conditions for the benefit of all living organisms. These concerns, combined with a growing understanding of renewable "green" resources, have prompted several sectors to take steps to reduce their harmful impact on the environment.

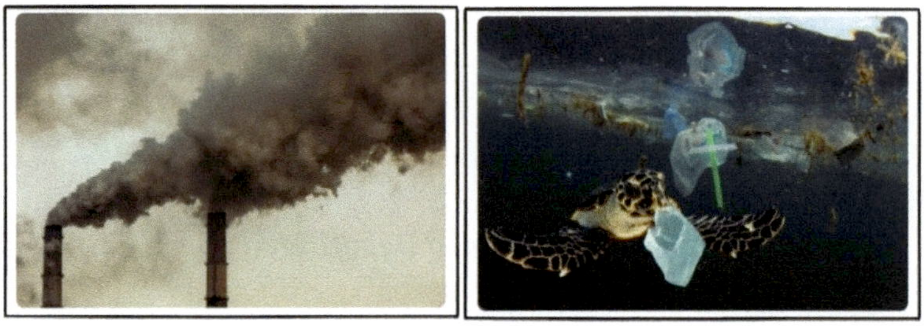

Figure 1.1 Environmental issues throughout the world.

In order to overcome the challenges, there is a demand for sustainably produced green materials with competitive abilities, with an emphasis to reduce the harmful impact on the environment. The global research on composites produced using green materials makes this area of research more enduring and powerful, with its contribution to reducing harmful impact on the environment using a sustainable approach. This chapter discusses the brief on the chapters, on composites and nanocomposites produced from green materials. The word "green" here refers to sustainable materials which are available in abundance and are left untouched as waste.

In recent years, sustainable development has become a major issue, and the looming depletion of oil-based resources will necessitate the usage of biomaterials derived from renewable resources. The widely accepted definition of sustainable development is "development that meets current demands without jeopardising future generation's ability to meet their own needs." In a broader sense, sustainable development is made up of three elements: society, the environment, and the economy. Hence, green composites are and form an essential part of sustainable development owing to their applications in different fields.

Biodegradation is the process of converting organic substrates (polymers) into low molecular weight fragments that may then be degraded to carbon dioxide and water using the activities of microbial species (Fig. 1.2). Biodegradation is influenced by a polymer's physical and chemical characteristics. The biodegradation mechanism of green composites and its evaluation methods are discussed in Chapter 2. Different standard testing methods used for biodegradation such as visual inspection, quantitative estimation of weight loss of the polymer, changes in properties such as a change in molecular weight, functional groups, crystallinity, tensile strength, measurement of CO_2 evolution or O_2 consumption, radiolabelling, and clear zone formation are discussed. Furthermore, the authors have discussed the biodegradation mechanisms of PLA (poly(lactic acid)), PHB, PHBV, and polymers extracted from biomass-based compounds [1, 2].

Figure 1.2 Process of biodegradation.

Cellulose is a polysaccharide made up of linear glucan chains linked by β-1,4 glycosidic bonds with cellobiose residues as the repeating units at various degrees of polymerization and packed into microfibrils held together by intramolecular hydrogen bonds and intermolecular van der Waals forces (Fig. 1.3). Green plants, algae, and oomycetes all have cellulose as a structural component of their major cell walls. Some bacteria emit it in order to create a biofilm. The most common organic polymer on the planet is cellulose. Cotton fiber has a cellulose percentage of 90%, wood has a cellulose level of 40–50%, and dry hemp has a cellulose content of around 57%. Chapter 3 discusses the green composites prepared by and/or using cellulose. Green composites made from cellulose offer a unique chance to develop new materials with improved socially acceptable features. According to the literature, these composites currently have sufficient characteristics to match glass fiber-based composites. Herein, the authors have elaborated on the different types of cellulose reinforcement (in terms of their morphology, production methods, and properties). In addition, thermoplastic, thermoset, and rubber-based composites in respect to their potential use together with micro- and nanostructured cellulosic fibers are also discussed [3, 4].

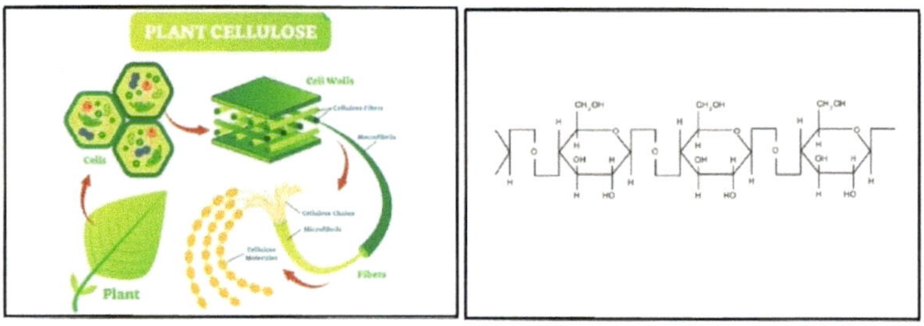

Figure 1.3 Cellulose source and chemical structure.

Chitin is a long-chain polymer made up of *N*-acetyl glucosamine, a glucose amide derivative. It is a key component of the cell walls of fungi, the exoskeleton of arthropods such as crustaceans and insects, and the radulae, cephalopod beaks, and gladii of mollusks and is the second most prevalent polysaccharide in nature (after only cellulose) (Fig. 1.4). At least some fish and Lissamphibia synthesize it as well. Chitin has a structure similar to cellulose, generating crystalline nanofibers or whiskers. It has a similar function to the protein keratin. Chitosan is a linear polysaccharide made up of *N*-acetyl-D-glucosamine (acetylated unit) and β(1→4)-linked D-glucosamine (deacetylated unit) that are randomly dispersed. Chitosan is manufactured by alkaline treatment of shrimp and other crustaceans' shells (Fig. 1.4). Chapter 4 discusses the green composites prepared by and/or using chitin and chitosan. Composites made by mixing fillers with biopolymers like chitin or chitosan are very promising materials since they improve characteristics while retaining biodegradability and non-toxicity. The use of hybrid polymer matrices comprising chitosan and other biodegradable polymers has been suggested to improve composite characteristics. Several research investigations reveal that chitosan may be molded into a variety of shapes, including hydrogels, films, powders, and particles, among others. The development of innovative chitosan-based polymeric materials is a fascinating way to investigate the applications of chitosan-based composites [5, 6].

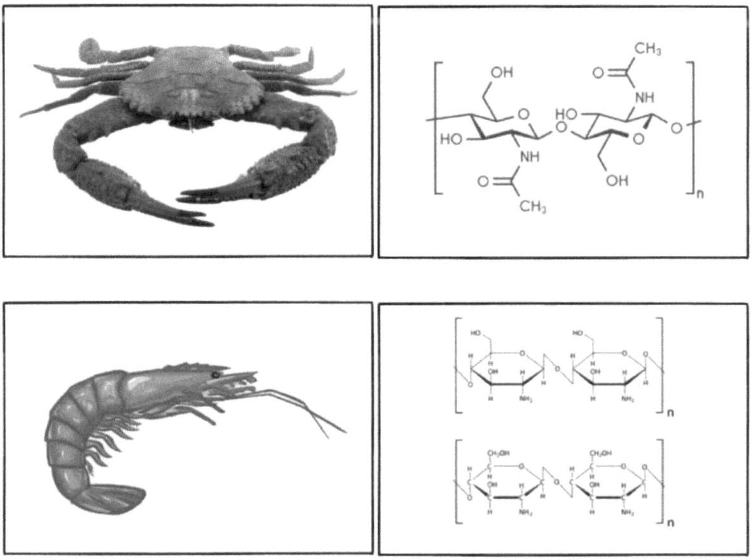

Figure 1.4 Chitin, chitosan source, and chemical structure.

Polyhydroxyalkanoates (PHAs) are polyesters created in nature by a variety of microorganisms, including bacteria that digest carbohydrates or lipids. They serve as a source of both energy and carbon storage when created by bacteria. Within this family, more than 150 distinct monomers can be mixed to create materials with a wide range of characteristics (Fig. 1.5). These biodegradable plastics are utilized in the manufacture of bioplastics. Thermoplastic and elastomeric materials with melting points ranging from 40°C to 180°C can be used. Blending, changing the surface, or combining PHAs with other polymers, enzymes, and inorganic elements can affect the mechanical properties and biocompatibility of PHAs allowing for a wider range of uses. Chapter 5 discusses the green composites prepared using PHAs. PHAs are biodegradable and biocompatible making them one of a kind composite materials with applications in biomedical research, packaging film, traditional plastics, membrane technology, and electronics. Wet and melt processing techniques can be used to create PHA micro- and nanocomposites. However, their production costs are now expensive, and their processing has several technological issues that must be resolved. Continued research into the issues connected with green composites based on PHAs could lower production costs and increase output volumes [7, 8].

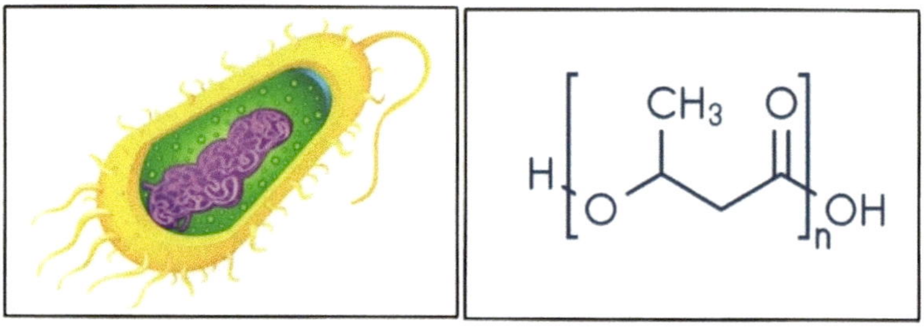

Figure 1.5 PHA source and chemical structure.

Poly(lactic acid) commonly known as PLA or poly(lactide) is a thermoplastic polyester made by condensing lactic acid with the removal of water (Fig. 1.6). Lactide is the cyclic dimer of the basic repeating unit and can also be produced through ring-opening polymerization. PLA has grown in popularity as a result of its cost-effective production from renewable resources. Its widespread use has been hampered by a number of physical and processing flaws. In 3D printing, PLA is the most extensively used plastic filament material. Chapter 6 discusses the green composites prepared using PLA. PLA is a biocompatible and biodegradable polymer that has a wide range of applications. PLA's use in a variety of industrial applications is predicted to increase dramatically in the next years, making it a cost-effective commodity plastic with the added virtue of being environmentally friendly. PLA is considered a "Green" polymer because it is produced in an environmentally friendly manner and can be used to replace petrochemical-based polymers [9, 10].

Figure 1.6 PLA source and chemical structure.

Protein-based composites provide a plethora of options for customizing material qualities by modifying the protein sequences used to make them or the methods used to make them. Chapter 7 discusses the green composites prepared using proteins. In this chapter, we look at a variety of different applications using a variety of proteins available from natural sources (Fig. 1.7). Proteins have shown promise in terms of intrinsic biodegradability, biocompatibility, natural abundance, and mechanical and functional qualities when compared to other natural polymers such as lignocellulosic components (cellulose, hemicellulose, and lignin) and starch. To summarize, the use of environmentally benign materials with improved properties provides a beneficial chance to partially replace petroleum-based plastics, while increasing the value of the huge natural macromolecules available [11, 12].

Figure 1.7 Protein sources.

Surface modification is the process of altering the surface of a material by adding physical, chemical, or biological features that differ from those found on the surface of the material originally (Fig. 1.8). Different approaches can be used to alter a wide range of surface attributes including roughness, hydrophilicity, surface charge, surface energy, biocompatibility, and reactivity. The current state of grafting approaches for surface modification of biobased polysaccharide nanoparticles, such as nanocelluloses, starch, and chitin nanoparticles, are discussed in Chapter 8 [13, 14].

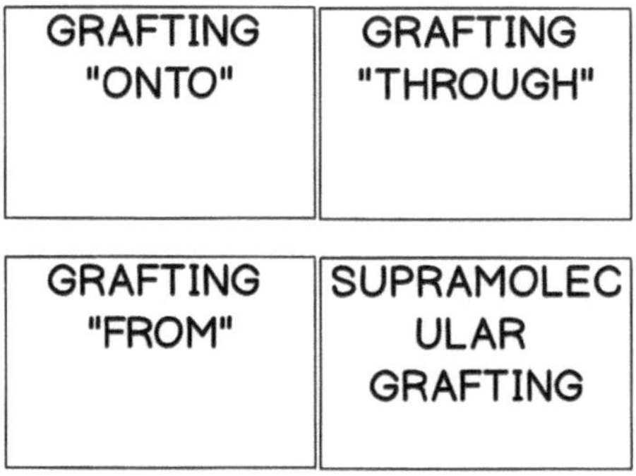

Figure 1.8 Grafting techniques.

The first section covers the various ways, such as grafting from, grafting onto, grafting through, and supramolecular grafting. The current state of the art in polysaccharide particle formation, physical and chemical properties, and relatively recent trends and methods are also updated. The final section delves into the most recent methodologies and successes in surface-modified nanocelluloses, nanochitin, and nanostarch particles via grafting from, grafting onto, and, most crucially, supramolecular grafting.

References

1. Wan YZ, Luo H, He F, et al. Mechanical, moisture absorption, and biodegradation behaviours of bacterial cellulose fibre-reinforced starch biocomposites. *Compos Sci Technol* 2009; 69: 1212–1217.
2. Kister G, Cassanas G, Bergounhon M, et al. Structural characterization and hydrolytic degradation of solid copolymers of D, L-lactide-co-ε-caprolactone by Raman spectroscopy. *Polymer (Guildf)* 2000; 41: 925–932.
3. Iwamoto S, Abe K, Yano H. The effect of hemicelluloses on wood pulp nanofibrillation and nanofiber network characteristics. *Biomacromolecules* 2008; 9: 1022–1026.
4. Alcalá M, González I, Boufi S, et al. All-cellulose composites from unbleached hardwood kraft pulp reinforced with nanofibrillated cellulose. *Cellulose* 2013; 20: 2909–2921.

5. Busila M, Musat V, Textor T, et al. Synthesis and characterization of antimicrobial textile finishing based on Ag:ZnO nanoparticles/chitosan biocomposites. *RSC Adv* 2015; 5: 21562–21571.
6. Rinaudo M. Chitin and chitosan: Properties and applications. *Prog Polym Sci* 2006; 31: 603–632.
7. Ma H, Liu M, Li S, et al. Application of polyhydroxyalkanoate (PHA) synthesis regulatory protein PhaR as a bio-surfactant and bactericidal agent. *J Biotechnol* 2013; 166: 34–41.
8. Barkoula NM, Garkhail SK, Peijs T. Biodegradable composites based on flax/polyhydroxybutyrate and its copolymer with hydroxyvalerate. *J Ind Crop Prod* 2010; 31: 34–42.
9. Rane A V., Kanny K, Mathew A, et al. Comparative analysis of processing techniques' effect on the strength of carbon black (N220)-filled poly (lactic acid) composites. *Strength Mater* 2019; 51: 476–489.
10. Rarima R, Unnikrishnan G. Poly(lactic acid)/gelatin foams by non-solvent induced phase separation for biomedical applications. *Polym Degrad Stab* 2020; 177: 109187.
11. Tanabe T, Okitsu N, Tachibana A, et al. Preparation and characterization of keratin-chitosan composite film. *Biomaterials* 2002; 23: 817–825.
12. Lim SW, Jung IK, Lee KH, et al. Structure and properties of biodegradable gluten/aliphatic polyester blends. *Eur Polym J* 1999; 35: 1875–1881.
13. Habibi Y. Key advances in the chemical modification of nanocelluloses. *Chem Soc Rev* 2014; 43: 1519–1542.
14. Malmström E, Carlmark A. Controlled grafting of cellulose fibres – an outlook beyond paper and cardboard. *Polym Chem* 2012; 3: 1702–1713.

Chapter 2

Biodegradability of Green Composites: Mechanisms and Evaluation Methods

Nikushi S. Yatigala,[a] Dilpreet S. Bajwa,[b]
Sreekala G. Bajwa,[c] and Saptaparni Chanda[b]
[a]*Agricultural and Biosystems Engineering, North Dakota State University, Fargo, ND 58108, USA*
[b]*Mechanical and Industrial Engineering, Montana State University, Bozeman, MT 59718, USA*
[c]*College of Agriculture, Montana State University, Bozeman, MT 59718, USA*
dilpreet.bajwa@montana.edu

Petroleum-based plastics cause substantial environmental problems at disposal because of their slow degradation and harmful degradation products. Most of the plastic wastes end up in landfills, creating water and soil pollution. For both water miscible and immiscible polymers that cannot be recycled nor incinerated, biodegradation is crucial as these polymers ultimately enter streams, which cannot be recycled nor incinerated. Biobased biodegradable polymers and natural fibers have a low environmental impact, and they are highly sustainable. They reduce waste accumulation,

Green Micro- and Nanocomposites
Edited by Sabu Thomas, Abitha V. K., and Hanna J. Maria

ISBN 978-981-4968-79-9 (Hardcover), 978-1-003-42756-8 (eBook)
www.jennystanford.com

control carbon dioxide emission, and minimize the dependency on petroleum-based fuels and products. The incorporation of natural fibers into biobased polymers causes an increase in specific strength and degradation rate and cost reduction. During biodegradation, polymers deteriorate their physical and chemical properties decreasing their molecular mass and form CO_2, H_2O, CH_4, and other low molecular weight products. This chapter discusses the current state of knowledge on the behavior of biobased polymers and their composites when subjected to accelerated weathering, ultraviolet, thermal, moisture, water, soil, and aerobic and anaerobic degradation. A range of biobased polymer composites are presented in this chapter, focusing on their degradation characteristics and mechanism, properties, physical and mechanical common standards, and evaluation methods.

2.1 Introduction

Low degradation rates and harmful degradation products such as carbon dioxide (CO_2) of synthetic polymers create substantial environmental problems at disposal. These polymers account for roughly 20% by volume of all wastes generated annually in the United States and contribute to global warming [1, 2]. Moreover, at the current rate of consumption, the petroleum that the polymers are created from can only last for another 50–60 years [3]. Owing to the negative impacts of petroleum-based plastics on the environment and ecosystem health, and the uncertain supply of fossil fuels, there is a great potential for replacing synthetic plastics with biodegradable alternatives. Not only do green plastics have a positive impact on the environment, but they also have many other attractive properties such as biocompatibility, bioactivity, chemical inertness, high stiffness and strength, good film-forming properties, and low toxicity [4–6]. For example, PLA, a chemically synthesized biobased polymer, is similar in mechanical properties to petrochemical polymers such as polypropylene and polyethylene [7]. Likewise, PHB, a polymer produced by bacteria, also has similar mechanical properties to polypropylene and higher moisture resistance and aroma barrier properties [8]. Similarly, cellulose, a naturally occurring polymer, is a hard polymer and has a high tensile strength [9]. When these biodegradable plastics are combined with biobased fibers and fillers to make green composites, their biodegradability generally improves.

Although green biodegradable polymers have numerous benefits, they are restricted in many applications as for their high cost and some of their undesirable properties such as low biodegradation rate, higher crystallinity, sensitivity to thermal degradation, and poor mechanical properties [5, 7, 10–12]. Compared to the waste accumulation rate, the degradation rate of biobased polymers such as PLA is still too slow [13]. The majority of biobased plastics tend to be brittle due to their high crystallinity. As for their thermal sensitivity, most green plastics are sensitive to thermal degradation during conventional melt processing. Reduction of molecular weight occurs as they degrade during thermal processing or under hydrolytic conditions. Mechanical properties of the processed biobased materials are likely to be affected by the reduction of molecular weight [14, 15]. However, mechanical and degradation properties can be changed while reducing cost by creating green composites with biobased polymers and other natural polymers or fillers [7, 12, 16]. Creating polymer blends and composites is a good cost-effective method to improve the properties such as dimensional stability, strength, toughness, and environmental degradation [7, 17]. In addition, mechanical properties of the biopolymer composites can be enhanced with reactive functional groups such as compatibilizers and coupling agents since they improve the adhesion between polymer and the fiber by forming both hydrogen and covalent bonds with hydroxyl groups of fiber and molecular entanglement with polymer [10, 16]. This chapter focuses on the biodegradation properties of biobased composites, how biopolymer composites containing various fiber and polymers behave, the different modes of biodegradation, how environmental conditions affect the biodegradation of these materials, and the standards for quantifying biodegradation properties.

2.2 Green Polymers

Biobased polymers serve as the matrix or substrate material in biocomposites, with natural fibers and other biobased materials serving as the fillers or reinforcement. It is important to understand the various types of bioplastics and their biodegradation characteristics in order to understand such properties of composites containing biopolymer as the substrate.

Polymers can be produced from various renewable and nonrenewable resources. Green or biobased polymers are the polymers produced by or from living systems. Some examples of green polymers from renewable resources are cellulose, starch, proteins, and DNA. Biodegradable plastics are polymers that will degrade due to the action of microorganisms, such as bacteria and fungi. Biodegradable polymers produced from renewable, biobased resources enjoy a high sustainability score. They reduce waste accumulation, control CO_2 emission, and lessen the need for petroleum-based fuels and products [18]. Bioplastics can be classified into biodegradable and nondegradable (Fig. 2.1). For example, biobased polyethylene is not biodegradable. Also, there are synthetic polymers (produced from nonrenewable petroleum resources) that are biodegradable [19]. As Fig. 2.2 illustrates, biodegradable green polymers from renewable resources can be divided into three categories: (i) those synthesized from bio-derived monomers (PLA), (ii) those produced by microorganisms (PHAs, bacterial cellulose), and (iii) those directly extracted from biomass with partial modification to meet the requirements [1, 6]. Table 2.1 provides some examples of naturally occurring polymers.

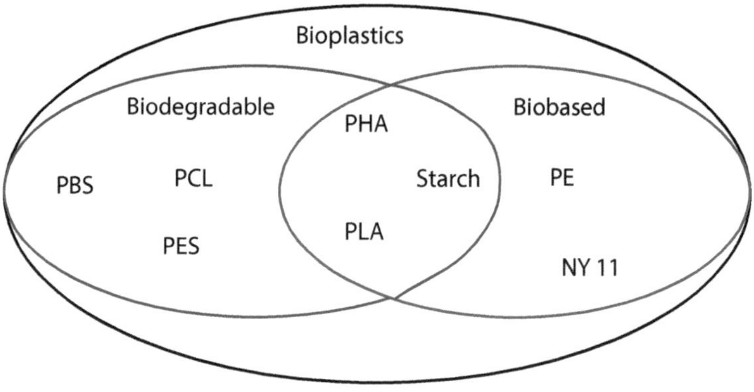

Figure 2.1 Types of bioplastics (adapted from [1]).

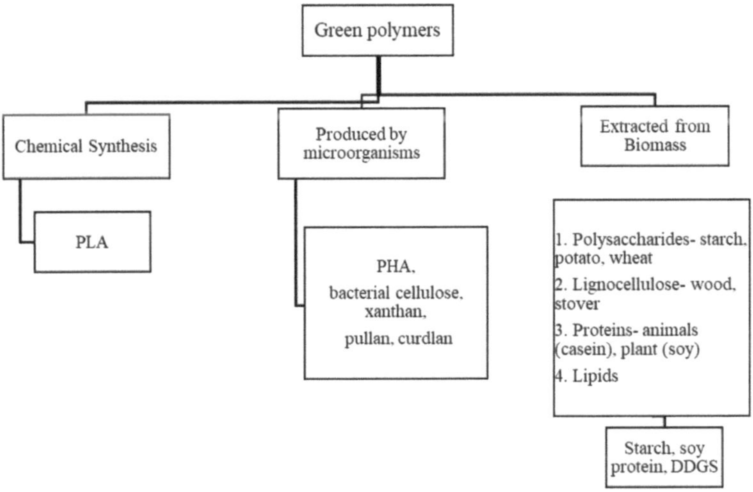

Figure 2.2 Origin of biobased polymers (adapted from [1]).

Table 2.1 Naturally occurring biodegradable polymers (adapted from [4])

Polyesters	**Polysaccharides (plant/algal)**
Polyhydroxyalkanoates	Starch (amylose/amylopectin)
Polylactic acid	Cellulose
Proteins	Agar
Silks	Alginate
Collagen/gelatin	Carrageenan
Elastin	Pectin
Resilin	Konjac
Adhesives	Various gums (e.g., guar)
Polyamino acids	**Polysaccharides (animal)**
Soy zein, wheat gluten	Chitin/chitosan
Casein, serum albumin	Hyaluronic acid
Polysaccharides (bacterial)	**Lipids/surfactants**
Xanthan	Acetoglycerides, waxes, surfactants
Dextran	Emulsan
Gellan	**Polyphenols**
Levan	Lignin

(*Continued*)

Table 2.1 (*Continued*)

Polyesters	Polysaccharides (plant/algal)
Curdlan	Tannin
Polygalactosamine	Humic acid
Cellulose (bacterial)	**Specialty polymers**
Polysaccharides (fungal)	Shellac
Pullulan	Poly-gamma-glutamic acid
Elsinan	Natural rubber
Yeast glucans	Synthetic polymers from natural fats
	Nylon from castor oil

2.3 Degradation of Biopolymers

There are many types of polymer degradation mechanisms: thermal, mechanical, photochemical, radiation chemical, biological, and chemical degradation [20]. According to ASTM definition, "degradable plastics are the plastics that are designed to undergo a significant change in their chemical structure under specific environmental conditions, resulting in a loss of some properties that may vary as measured by standard methods appropriate to the plastic and the application in a period of time that determines its classification." This definition can be applied to many polymer degradation types including photodegradation, thermooxidation, hydrolysis, and biodegradation. Autooxidation is the cause of thermooxidative degradation of organic materials without light. In natural photodegradation, the molecular weight of the polymers is decreased by sunlight due to photooxidation and direct bond cleavage [21]. Another mode of degradation is hygrothermal degradation. Substantial loss of weight and mechanical properties of a material due to the effects of moisture and temperature is called hygrothermal degradation [22]. In nature, to decompose organic matter, biotic and abiotic factors act synergistically. Some biodegradation studies show that abiotic degradation precedes microbial assimilation [23–25]. Exposing polymers to naturally degrading environmental conditions (weather, aging, and burying) results in different transformations of polymers due to light, thermal, and chemical agents. These abiotic parameters are useful factors in initiating the biodegradation process, as these parameters help to weaken the polymer structures [25, 26]. Some other mechanisms of polymer degradation include the following:

(i) Chemical degradation: It occurs when the polymers are brought to contact with chemicals such as acids, bases, and solvents. Oxygen is the most important as it attacks covalent bonds creating free radicals [20].

(ii) Thermal degradation: It is the molecular deterioration of polymer due to heat. For thermoplastics, this occurs at the melting temperature. The components of the polymer backbone chain detach (molecular scission) and react with each other resulting in changed polymer properties.

(iii) Mechanical degradation: Macroscopic changes occur in polymer material due to compression, tension, and shear forces. Under shear forces, molecular interaction between certain molecules at certain sites gets disrupted.

(iv) Photodegradation: In photodegradation, polymers undergo physical and chemical changes due to ultraviolet light or visible light, inducing cross-linking/Norrish reactions, or oxidative processes.

2.4 Biodegradation

During biodegradation, living organisms break down organic substances, and thus the chemical structure of the material changes resulting in the production of carbon dioxide, water, and new microbial cell mass. Only biodegradation or biological degradation completely removes the polymer or its degradation products from the environment. It takes about 1–3 years to completely assimilate and disappear a biodegradable plastic article [21]. Biodegradation can occur under two different conditions: aerobic and anaerobic. Polymers can be attacked by living organisms either chemically or mechanically. Enzymes are involved in the chemical mode while the mechanical mode refers to the attack by mammals and insects [20]. Enzymes are proteins of complicated chemical structures. Typically, the catalytic activity of enzymes is related to a special molecular conformation. Depending on the properties of the biopolymer, the biodegradation of plastics proceeds under different soil conditions actively due to ideal growth conditions of microorganisms in the soil [15].

In biodegradation, microbes such as bacteria and fungi are engaged in the degradation of both bio- and petroleum-based plastics. Bacteria involved in the biodegradation process include, *inter alia*, *Bacillus*, *Pseudomonas*, *Klebsiella*, *Actinomycetes*, *Nocardia*, *Streptomyces*, *Thermoactinomycetes*, *Micromonospora*, *Mycobacterium*, *Rhodococcus*, *Flavobacterium*, *Comamonas*, *Escherichia*, *Azotobacter*, and *Alcaligenes* (some of them are capable of consuming polymer about 90% of their dry mass). Fungi active in the biodegradation process are *Sporotrichum*, *Talaromyces*, *Phanerochaete*, *Ganoderma*, *Thermoascus*, *Thielavia*, *Paecilomyces*, *Thermomyces*, *Geotrichum*, *Cladosporium*, *Phlebia*, *Trametes*, *Candida*, *Penicillium*, *Chaetomium*, and *Aerobasidium* [27].

Biodegradation depends on many factors such as the type of organism, pretreatment method, polymer characteristics such as mobility, molecular weight, crystallinity, and plasticizers or additives blended with the polymer. For instance, higher molecular weight lowers the solubility of the polymer, reducing the microbial attack, and the rate of degradability [15]. The insolubility of polymers reduces the availability of digestible substrates through bacterial cellular membrane.

As opposed to degradable/biodegradable synthetic plastics, biodegradable green polymers/composites generally do not produce toxic products, or it is much less compared to synthetic plastics. Biodegradable green composites are not necessarily easy to decompose, and they require certain environmental conditions. Even then, they may leave behind toxic residues. There are three types of degradable plastics:

(1) Degradable: Plastics will undergo a significant change in their chemical structure under certain environmental conditions resulting in a loss of some properties.
(2) Biodegradable: Plastics will degrade due to microbial activity producing CO_2, water, and biomass as end products. There is no requirement for leaving "no toxic residue," nor for the time it needs to take to biodegrade. Thus, "biodegradable"

green polymers/composites can leave toxic residues during biodegradation.

(3) Compostable: According to ASTM, compostable plastic should:

(i) Biodegrade – break down into carbon dioxide, water, and biomass at the same rate as cellulose.

(ii) Disintegrate – the material is indistinguishable in the compost.

(iii) Not produce any toxic material and the compost can support plant growth.

2.4.1 Biodegradation Mechanism Steps

During degradation, polymers initially break down into smaller, simpler structures such as oligomers and monomers (Fig. 2.3). These smaller structures of monomers are mineralized to CO_2, H_2O, or CH_4 as end products [15]. Biodegradability is the complete assimilation of the degraded products as a food source by microorganisms, and it ensures the effective and safe returning of carbon into the ecosystem. In addition, at least two categories of cellular enzymes also participate in the biodegradation process of polymers: extracellular and intracellular depolymerases. The initial breakdown of the polymers prompts different kinds of physical forces to occur such as heating/cooling, freezing/thawing, and wetting/drying, which result in mechanical damage in the polymer such as cracking [15]. Biodegradation of polymers includes the following steps, and the process could stop at any stage [25].

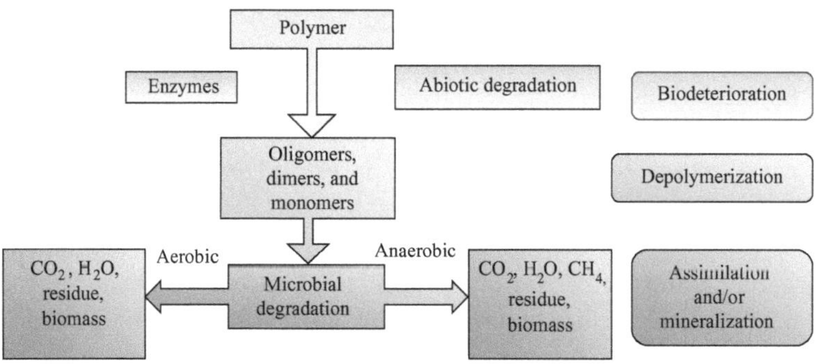

Figure 2.3 General mechanism of polymer biodegradation.

(a) Biodeterioration

Microbial or/and abiotic factors break the biodegradable materials into small fractions. In thermoplastics, depending on the rate of chemical diffusion to the materials, and the rate of polymer bond cleavage, biodeterioration occurs in two ways: bulk and surface erosion (Fig. 2.4). In bulk erosion (when the chemical diffusion rate is faster than polymer bond breakage), chemicals or radiation activity induces bond cleavage resulting in changes in molecular weight of the polymer matrix. On the other hand, in surface erosion (when the polymer bond breakage rate is faster than the chemical diffusion rate), even though the loss of substances occurs, the molecular weight of the polymer matrix does not change.

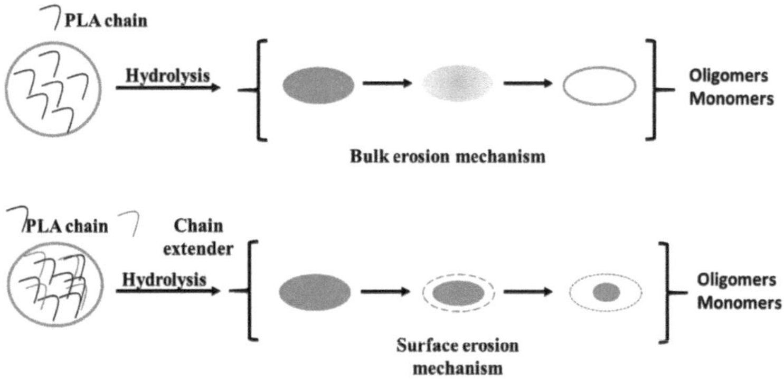

Figure 2.4 Bulk and surface erosion mechanisms of PLA.

Microorganisms (bacteria, protozoa, algae, fungi, and lichenaceae groups) that develop on the surface and/or inside the material act by physical, enzymatic, and/or chemical means:

Physical – Polymers such as polysaccharides and proteins discharge a glue textured substance, which is a complex matrix. This slime matrix changes the size and distribution of pores, and the degree of moisture and thermal transfers. Also, this substance helps microorganisms to stick to the material surfaces, protecting microorganisms against undesirable conditions such as UV radiation. Penetration of microorganisms into the materials results in increases in the size of pores and cracks, thus weakening the durability of the material.

Enzymatic – For enzymes to break down specific bonds, the presence of factors such as cations in the material matrix and coenzymes synthesized by microorganisms are important.

Chemical – Microbial degradation produces extracellular polymers that act as surfactants that assist the interactions between hydrophilic and hydrophobic phases, increasing the microbial penetration rate. Each of these microorganisms contributes to the chemical biodeterioration of the material. For instance, chemoorganotrophic microorganisms use organic substrates as carbon, energy, and electron sources. They release organic acids, such as oxalic, citric, gluconic, glutaric, glyoxalic, oxaloacetic, and fumaric acids. Polymers such as PLA releases succinic acid, adipic acid, lactic acid, and butanediol due to abiotic and/or biotic hydrolysis. Hydrolysis, which leads to generating oligomers and monomers, is initiated when water enters the polymer matrix (Fig. 2.5).

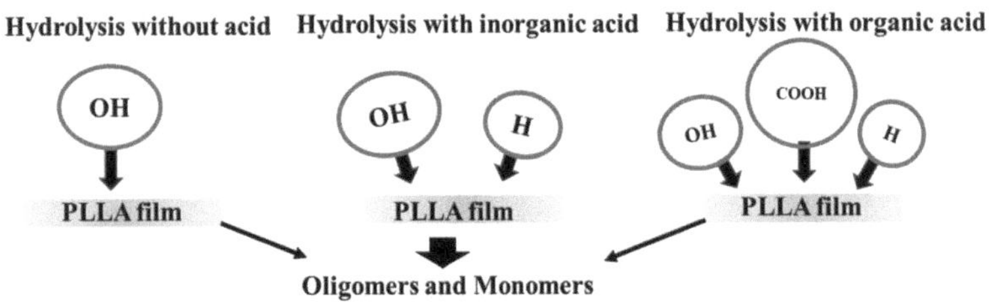

Figure 2.5 PLA hydrolysis in acidic conditions.

(b) Depolymerization

Microorganisms produce catalytic agents (enzymes and free radicals) that could split polymeric molecules resulting in molecular weight reduction. This process generates oligomers, dimers, and monomers, and these are small in size to transfer through the semipermeable outer bacterial membranes (Fig. 2.6). These products are then exploited as carbon and energy sources [15]. The molecules that are recognizable by microbial cell receptors could go through the plasmic membrane, while the other molecules remain in the extracellular surroundings. The remaining molecules are subjected to biotransformation reactions producing assimilable or unassimilable products.

Polymer scissions are essential to obtain oligomers and/or monomers. The energy to achieve scissions can be acquired from different sources: thermal, light, mechanical, chemical, and/or biological. In biological, microorganisms produce specific enzymes or free radicals. Enzymes prompt chemical reactions by reducing the activation energy level of molecules. Discharged enzymes (to the extracellular environment) can be found as free catalysts or fixed on particles.

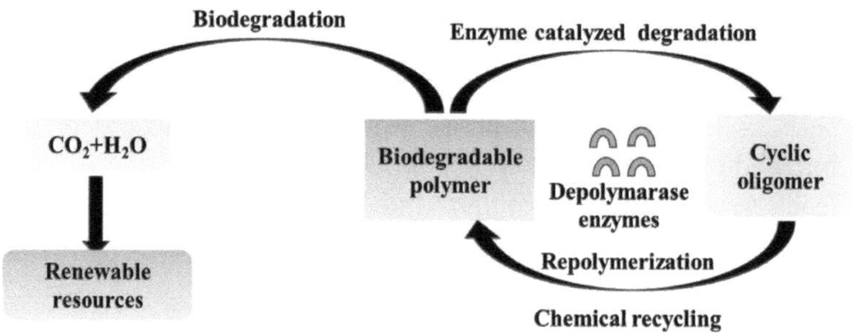

Figure 2.6 Biodegradation of polymers by enzymes.

(c) Assimilation

Transferred molecules integrate the microbial metabolism in the cytoplasm in order to produce energy, new biomass, storage vesicles, and many types of metabolites that help in maintaining cellular activity, structure, and reproduction. Thus, microorganisms grow while reproducing and consuming nutrient substrate from the environment. Three different catabolic pathways exist in producing the energy depending on aerobic or anaerobic conditions: aerobic respiration, anaerobic respiration, and fermentation.

Aerobic respiration – Microorganisms that can use oxygen as the final electron acceptor need substrates that are oxidized into the cell. Initially, basic catabolic pathways such as glycolysis, β-oxidation, amino acid catabolic reactions, and purine and pyrimidine catabolism generate a small amount of energy. Oxidative phosphorylation that is realized by electron transport systems produces more energy afterward.

Anaerobic respiration – Microorganisms that cannot use oxygen as the final electron acceptor initiate total oxidation with adapted electron transport in membrane systems using NO, SO_2, S, CO, Fe^{3+}, and fumarate as final electron acceptors.

Fermentation – Microorganisms that are unable to use oxygen or other exogenous mineral molecules as final electron acceptors produce energy by fermentation. These microbes use self-synthesized endogenous organic molecules as final electron acceptors. Mineral and/or organic molecules discharged into the environment such as CO_2, ethanol, lactate, acetate, and butanediol are the products of fermentation.

(d) Mineralization

At the same time, it is possible for particular metabolites to reach the extracellular surroundings. Molecules such as CO_2, N_2, CH_4, H_2O, and different salts from intracellular metabolites are released into the environment. When O_2 is available, mostly aerobic microorganisms destroy the complex materials producing microbial biomass, CO_2, and H_2O as final products. Under anoxic conditions, anaerobic microorganisms deteriorate the polymer. Microbial biomass, CO_2, CH_4, and H_2O are the end products [15]. Consequently, polymers biodegrade under two conditions: aerobic (in the presence of oxygen) and anaerobic (in the absence of oxygen).

2.5 Standard Testing Methods of Biodegradation

Biodegradability of the polymers is evaluated by measuring the weight loss of degraded material, reduction in mechanical properties, change in molecular weight, quantitative estimation of microbial growth, visual observations, and/or monitoring the amount of the degradation products that may include the evolution of CO_2 and consumption of O_2 [28]. Table 2.2 includes the standards that are used in evaluating the biodegradability of polymers.

Some of the methods of examining biodegradation are described below.

1. Visual inspection of mycelium growth on the polymer surface

Visually observable changes are roughening of the surface, holes or cracks formation, defragmentation, microorganism development on the surface, and changes in the color of the composites [25]. Crystalline spherulites can be seen on the surface after initial degradation-biodeterioration. It is the degradation of amorphous polymer fraction, scraping the less degradable crystalline parts from the material. Even though these changes do not prove the biodegradation of the polymers in terms of metabolism, they indicate the attacks of microbes [15]. Microscopic techniques such as photonic microscopy, electronic microscopy, and polarization microscopy are used to refine the analysis. More detailed observations can be made using scanning electron microscopy (SEM) and atomic force microscopy (AFM). Standards such as ASTM G21-70, ASTM G22–76, ISO 846, NF X41-514, NF X41-513, and ISO 11266 test the biodegradation by microorganism colonization on Petri dishes.

2. Quantitative estimation of the weight loss of the polymer

Even though this method is commonly used, weight loss of samples is not a direct proof of a material biodegradability as it can be due to the disappearance of volatile and soluble impurities, and loss of material from the sample. However, placing samples in small nets can facilitate the recovery. Nonetheless, detailed information on biodegradation can be acquired by structural analysis of both the material remains and the low molecular weight intermediates [15]. Typical standards used are NF EN ISO 13432, ISO 14852, and ISO 14855.

3. Measurement of changes in polymer properties, such as changes in molecular weight, functional groups, crystallinity, tensile strength, or a combination thereof

Changes in mechanical properties do not also provide direct proof of biodegradation. However, these variations are considered when changes in the mass of the samples are observed. The tensile strength of polymers is highly sensitive to the changes in the molar mass of polymers, which is often taken as a direct indication of degradation [15]. Determinations of molecular weight (MW) and molecular weight distribution (MWD) are strong methods of detecting the degradation of linear soluble polymers. Random main scission, a chemical reaction in the main chains or at side groups of linear polymers, results in a change in average MW. MWD is altered upon main chain rupture and/or cross-linking. MWD can be determined by gel permeation chromatography (GPC) [20]. Thermal evolution is studied in many biodegradation studies using a differential scanning calorimetry (DSC). This provides the glass transition temperature (T_g), cold crystallization temperature (T_{cc}), and melting temperature (T_m). X-ray diffraction (XRD) is used to determine crystallinity [25]. The formation and disappearance of chromophoric groups are detected by Fourier transform infrared spectroscopy (FTIR), while nuclear magnetic resonance spectroscopy (NMR) analyzes structural changes [20]. There are other techniques that can be used in assessing biodegradability such as X-ray photoelectron spectroscopy (XPS), contact angle measurements, and water uptake [15].

4. Measurement of CO_2 evolution/O_2 consumption

Measuring the amount of oxygen consumed (respirometric test – ISO 14852) or the amount of CO_2 formed (Sturm test) is a more accurate way of evaluating polymer degradation, as microorganisms utilize O_2 to oxidize carbon and form CO_2 under aerobic conditions. There are a few methods used in detecting O_2 and CO_2 concentrations such as trapping of CO_2 in $Ba(OH)_2$ solution, monitoring O_2 and CO_2 concentrations in the air by infrared and paramagnetic O_2 detectors, detecting decreased O_2 by the diminution of the pressure, and utilizing small sealed bottles as degradation reactors in which the CO_2 in the headspace and the drop in dissolved oxygen are dissected [15]. On the other hand, in anaerobic conditions, augmentation of the pressure is measured when gases are released. Evolved gasses are identified by using gas chromatography [25].

CO_2 analysis is also used for tests in solid matrices such as in controlled composting test (ASTM 5338, DIN V 54900, ISO 14855, JIS K 6953). The only efficient method for analyzing CO_2 in a complex environment is the use of a radiolabeled polymer to perform CO_2 respirometry. Unfortunately, in addition to its hazardousness, it is a time-consuming and expensive test, since it requires a specific lab room with specific equipment with special training technicians.

5. Radiolabeling

Radiolabeling is a consistent and expensive method of estimating the evolved amount of ^{14}C carbon dioxide. Using a scintillation counter, ^{14}C carbon dioxide evolution can be determined by exposing materials with randomly distributed ^{14}C markers to microbial environments [15].

6. Clear-zone formation

This method is used to determine the ability of microorganisms to depolymerize the polymer. Here, the polymer is scattered as fine particles among the synthetic medium agar, resulting in the agar with a murky appearance. Following vaccination with microorganisms, if a clear halo is developed around the settlement, it means that the microorganisms could depolymerize the polymer. Ordinarily, this system is used to screen microbes that have the ability to degrade specific polymer; however, by investigating the development of clear zones, it can likewise be utilized to acquire semiquantitative outcomes [15].

Table 2.2 Standards for practicing and testing biodegradability of polymers (adapted from [29–31])

Standard	Composting
ASTM D5209	Standard Test Method for Determining the Aerobic Biodegradation of Plastic Materials in the Presence of Municipal Sewage Sludge
ASTM D5338	Standard Test Method for Determining the Aerobic Biodegradation of Plastic Materials Under Controlled Composting Conditions
ASTM D5509	Standard Practice for Exposing Plastics to a Simulated Compost Environment
ASTM D5512	Standard Practice for Exposing Plastics to a Simulated Compost Environment Using an Externally Heated Reactor
ASTM D6003	Standard Test Method for Determining Weight Loss from Plastic Materials Exposed to a Simulated Municipal Solid Waste (MSW) Aerobic Compost Environment
ASTM D5988	Standard Test Method for Determining the Aerobic Biodegradation in Soil of Plastic Materials or Residual Plastic Materials After Composting
ASTM D6002	Guide to Assess the Compostability of Environmentally Degradable Plastics – ISR Program

Standard	Composting
ASTM D6340	Standard Test for Determining Aerobic Biodegradation of Radiolabeled Plastic Materials in Compost Environment – ISR Program
ASTM D6400	Specifications for Compostable Plastics – ISR Program
ASTM D5951	Standard Practice for Preparing Residual Solids Obtained After Biodegradability Standard Methods for Toxicity and Compost Quality Testing – Fate & Effect Testing
ASTM D5152	Standard Practice for Water Extraction of Residual Solids from Degraded Plastics for Toxicity Testing – Fate & Effect Testing
ASTM D6868	Standard Specification for Biodegradable Plastics Used as Coatings on Paper and Other Compostable Substrates
ASTM G160	Standard Practice for Evaluating Microbial Susceptibility of Nonmetallic Materials by Laboratory Soil Burial
ASTM D6954	Standard Guide for Exposing and Testing Plastics That Degrade in the Environment by a Combination of Oxidation and Biodegradation
ASTM G29	Standard Practice for Determining Algal Resistance of Plastic Films
EN ISO 846	Plastics – Evaluation of the action of microorganisms
ISO 846	Plastics: Determination of behavior under the action of fungi and bacteria. Evaluation by visual examination or measurement of changes in mass or physical properties
OECD 304 A	Inherent Biodegradability in Soil
	Anaerobic digestion/processes
ASTM D5988	Standard Test Method for Determining the Aerobic Biodegradation of Plastic Materials or Residual Plastic Materials After Composting in Contact with Soil

(*Continued*)

Table 2.2 (*Continued*)

Standard	Anaerobic digestion/processes
ASTM D5271	Standard Test Method for Assessing the Aerobic Biodegradation of Plastic Materials in an Activated-Sludge-Wastewater-Treatment System
ASTM D5210	Standard Test Method for Determining the Anaerobic Biodegradation of Plastic Materials in the Presence of Municipal Sewage Sludge
ASTM D5511	Standard Test Method for Determining Anaerobic Biodegradation of Plastic Materials Under High-Solids Anaerobic Digestion Conditions
ASTM D5526	Standard Test Method for Determining Anaerobic Biodegradation of Plastic Materials Under Accelerated Landfill Conditions
ASTM D5525	Standard Practice for Exposing Plastics to a Simulated Landfill Environment
	Others
ASTM D5437	Standard Practice for Weathering of Plastics Under Marine Floating Exposure
ASTM D5510	Standard Practice for Heat Aging of Oxidatively Degradable Plastics
ASTM D6118	Standard Test Method for Determining Hydrolytic Degradation of Plastic Materials in an Aqueous Solution
ASTM D1435	Standard Practice for Outdoor Weathering of Plastics
ASTM D4364	Standard Practice for Performing Outdoor Accelerated Weathering Tests of Plastics Using Concentrated Sunlight
OECD 306	Biodegradability in Seawater
ASTM D5247	Standard Test Method for Determining the Aerobic Biodegradability of Degradable Plastics by Specific Microorganisms
ASTM D5209	Standard Test Method for Determining the Aerobic Biodegradation of Plastic Materials in the Presence of Municipal Sewage Sludge
OECD 303 A	Aerobic Sewage Treatment: Activated Sludge Units

Standard	Others
ASTM D5071	Practice for Operating Xenon Arc-Type Exposure Apparatus with Water for Exposure of Photodegradable Plastics
ASTM D5208	Practice for Operating Fluorescent UV and Condensation Apparatus for Exposure of Photodegradable Plastics
ASTM D5272	Standard Practice for Outdoor Exposure Testing of Photodegradable Plastics
	Evaluation by the content of carbon, oxygen, CO_2, biogas, etc.
OECD 301 A	DOC Die-Away Test
OECD 301 B	CO_2 Evolution Test
OECD 301 C	Modified MITI Test
OECD 301 D	Closed Bottle Test
OECD 301 E	Modified OECD Screening Test
OECD 301 F	Manometric Respirometry Test
OECD 302 A	Modified SCAS Test
OECD 302 B	Zahn-Wellens Test
OECD 302 C	Modified MITI Test
ISO 7827	Water quality – Evaluation in an aqueous medium of the "ultimate" aerobic biodegradability of organic compounds – Method by analysis of dissolved organic carbon (DOC)
ISO 9439	Water quality – Evaluation in an aqueous medium of the "ultimate" aerobic biodegradability of organic compounds – Method by analysis of released carbon dioxide
ISO 9408	Water quality – Evaluation in an aqueous medium of the "ultimate" aerobic biodegradability of organic compounds – Method by determining the oxygen demand in a closed respirometer
ISO 9887	Water quality – Evaluation of the aerobic biodegradability of organic compounds in an aqueous medium – Semi-continuous activated sludge method (SCAS)

(*Continued*)

Table 2.2 (*Continued*)

Standard	Evaluation by the content of carbon, oxygen, CO_2, biogas, etc
ISO 9888 B	Water quality – Evaluation of the aerobic biodegradability of organic compounds in an aqueous medium – Static test (Zahn-Wellens method)
ISO 10634	Water quality – Guidance for the preparation and treatment of poorly water-soluble organic compounds for the subsequent evaluation of their biodegradability in an aqueous medium
ISO 10707	Water quality – Evaluation in an aqueous medium of the ultimate aerobic biodegradability of organic compounds – Method by analysis of biochemical oxygen demand (closed bottle test)
ISO 10708	Water quality – Evaluation in an aqueous medium of the ultimate aerobic biodegradability of organic compounds – Method by determining the biochemical oxygen demand in a two-phase closed bottle test
ISO 11733	Water quality – Evaluation of the elimination and the biodegradability of organic compounds in an aqueous medium. Activated sludge simulation test
ISO 11734	Water quality – Evaluation of the ultimate anaerobic biodegradability of organic compounds in digested sludge. Method by measurement of the biogas production
ISO/FDIS 14592-1	Water quality – Evaluation of the aerobic biodegradability of organic compounds at low concentrations – Part 1: Shake-flask batch test with surface water or surface water/sediment suspension
ISO/DIS 14592-2	Water quality – Evaluation of the aerobic biodegradability of organic compounds at low concentrations – Part 2: Continuous flow river model with attached biomass
ISO 14593	Water quality – Evaluation of ultimate aerobic biodegradability of organic compounds in aqueous medium – Method by analysis of inorganic carbon in sealed vessels (CO_2 headspace test)
ISO/TR 15462	Water quality – Selection of tests for biodegradability

Standard	Evaluation by the content of carbon, oxygen, CO_2, biogas, etc
ISO 16221 B	Water quality – Guidance for determination of biodegradability in the marine environment
EN ISO 7827	Water quality – Evaluation in an aqueous medium of the "ultimate" aerobic biodegradability of organic compounds – Method by analysis of dissolved organic carbon (DOC)
EN ISO 9439	Water quality – Evaluation of ultimate aerobic biodegradability of organic compounds in aqueous medium – Carbon dioxide evolution test
EN ISO 9408	Water quality – Evaluation of ultimate aerobic biodegradability of organic compounds in aqueous medium by determination of oxygen demand in a closed respirometer
EN ISO 9887	Water quality – Evaluation of the aerobic biodegradability of organic compounds in an aqueous medium – Semi-continuous activated sludge method (SCAS)
EN ISO 9888	Water quality – Evaluation of ultimate aerobic biodegradability of organic compounds in aqueous medium – Static test (Zahn-Wellens method)
EN ISO 10634	Water quality – Guidance for the preparation and treatment of poorly water-soluble organic compounds for the subsequent evaluation of their biodegradability in an aqueous medium
EN ISO 10707	Water quality – Evaluation in an aqueous medium of the "ultimate" aerobic biodegradability of organic compounds – Method by analysis of biochemical oxygen demand (closed bottle test)
EN ISO 11733	Water quality – Evaluation of the elimination and biodegradability of organic compounds in an aqueous medium – Activated sludge simulation test
EN ISO 11734	Water quality – Evaluation of the "ultimate"; anaerobic biodegradability of organic compounds in digested sludge – Method by measurement of the biogas production
ISO 8192 B	Water quality – Test for inhibition of oxygen consumption by activated sludge

(Continued)

Table 2.2 (*Continued*)

Standard	Evaluation by the content of carbon, oxygen, CO_2, biogas, etc
ISO 9509 B	Water quality – Method for assessing the inhibition of nitrification of activated sludge microorganisms by chemicals and waste waters
ISO 10712 B	Water quality – Pseudomonas putida growth inhibition test (Pseudomonas cell multiplication inhibition test)
ISO 11348 Part 1, 2, 3	Water quality – Determination of the inhibitory effect of water samples and the light emission of Vibrio fischeri (Luminescent bacteria test)
EN ISO 8192	Water quality – Test for inhibition of oxygen consumption by activated sludge
EN ISO 9509	Water quality – Method for assessing the inhibition of nitrification of activated sludge microorganisms by chemicals and waste waters
EN ISO 10712	Water quality – Pseudomonas putida growth inhibition test (Pseudomonas cell multiplication inhibition test)
EN ISO 11348 Teil 1, 2, 3	Water quality – Determination of the inhibitory effect of water samples on the light emission of Vibrio fischeri (Luminescent bacteria test)

2.6 Biodegradation Properties of Biopolymers and Composites

Polymer composites can be formulated to have specific mechanical and physical properties that do not exist in nature, and this increases the range of applications that these biomaterials can be used [18]. Polymer composites can be designed for specific mechanical performance through a selection of the polymer and natural fibers, their formulation, and processing methods [9, 32]. Polymers such as PLA, PHA, and starch are blended with other natural polymers and fibers to increase and control biodegradation [9, 12, 33]. In order to use these biodegradable composites for outdoor and indoor applications, it is important to understand the degradation behavior of the composites.

Natural fibers used in biodegradable composites are obtained either from plants or animals. Compared to commonly used glass fiber that is not biodegradable, the mechanical properties of natural fibers are inferior. However, natural fibers have low density and, therefore, have a high strength-to-density ratio and stiffness [19, 34]. Fibers are commonly used as reinforcement or fillers in biocomposites for many advantages such as their low cost, low density, high specific strength, low health hazards during processing and utilization; good thermal, electric, and acoustic insulation properties; and less abrasiveness to processing equipment as compared to conventional inorganic fillers [35]. Even though natural fibers offer many benefits, there are some factors that should be taken into consideration. Poor adhesion between the fiber and the matrix, poor moisture resistance, flame-retardant properties, homogenization of the fiber's properties, degree of polymerization and crystallization, aggregation during processing, and lower durability are some of the challenges [14]. The mechanical properties of composites strongly depend on the quality of the fiber-matrix interface [35]. The main reason for poor interfacial adhesion between the matrix and fibers is the hydroxyl and other polar groups of natural fibers which make them hydrophilic [36]. Hemicelluloses and lignin in natural fiber are amorphous with a high affinity for water. The hydrophilic fractions of the fibers with their free hydroxyl group make it incompatible with the hydrophobic polymer matrix [14]. This leads to the poor mechanical performance of the composites [15, 37]. However, interfacial treatments can improve the bonding between the fiber and the polymer. Many studies have reported the changes exhibited by green composites and explained their degradation characteristics and changes in mechanical properties. A summary of degradation studies conducted on biobased polymers is given in Table 2.3.

Many biobased composites use nanosized mineral fillers such as clay. Recent studies indicate that layered silicate-filled polymer composites exhibit improved mechanical, thermal, and physicochemical properties at low filler concentrations, primarily due to the nano-level interactions with the polymer matrix. Montmorillonite (MMT), hectorite, and saponite are the commonly used clays for polymer composites [38]. Quaternary alkylammonium salt is a commonly used biodegradable (under certain waste treatment conditions) organic modifier with nanoclays [38].

However, some studies show a low biodegradation because of the superior barrier properties of nanoclay composites [39], while some do not show any changes in 60-day biodegradation between pristine biopolymers and their nanoclay composites [40].

Table 2.3 Summary of degradation studies conducted on biodegradable biobased polymers and composites

Title	Manufacturing procedure	Results	Reference
Effects of aging on the thermomechanical properties of poly(lactic acid)	Melt mixed using Brabender mixer and then compression molded. Samples were exposed to an RH of 80% at different temperatures (20°C, 40°C, and 50°C) for up to 130 days	Higher degradation at higher temperatures	Niaounakis et al. [13]
Thermal degradation and physical aging of poly(lactic acid) and its blends with starch	Melt mixed using twin-screw extruder and then compression molded. Stored the samples at 25°C with humidity between 30% and 90%	Thermal endurance of PLA/starch/MDI was greater than PLA/starch, but lower than neat PLA	Acioli-Moura et al. [44]
Evaluation of PLA–lignin bioplastics properties before and after accelerated weathering	Melt mixed using Brabender mixer and then compression molded. Samples were exposed to accelerated weathering up to 600 h	Tensile and impact strength decreased for all samples. Increased water sorption for all composites	Spiridon et al. [32]
Preparation, characterization, and biodegradability of renewable resource-based composites from recycled polylactide bioplastic and sisal fibers	Melt mixed using plastograph mixer and compression molded. Samples were soil buried	Lower biodegradability rate for PLA-g-AA/SF than PLA/SF composites, but higher than neat PLA	Wu [37]

Title	Manufacturing procedure	Results	Reference
Renewable resource-based composites of recycled natural fibers and malleated polylactide bioplastic: characterization and biodegradability	Melt mixed and then compression molded. Samples were exposed to *Burkholderia cepacia* bacterium	Lower biodegradability rate for PLA-g-MA/GCF than PLA/GCF, but higher than PLA. Higher degradation rate with increased fiber content	Wu [33]
Characteristics of a silk fiber reinforced biodegradable plastic	Extruded and then injection molded. Samples were stored in phosphate buffered saline (PBS) solution tanks	Higher water absorption, resulting in higher biodegradability	Ho et al. [45]
Biodegradation of a silkworm silk/PLA composite	Extruded and injection molded. A biodegradation test was carried out using PBS solution	Higher biodegradation rate for PLA/silk than neat PLA	Cheung et al. [46]
Long-term immersion in natural seawater of Flax/PLA biocomposite	Extruded and then injection molded. Aged in natural seawater for 2 years	A linear relationship between water uptake and loss of mechanical properties	Duigou et al. [47]
Moisture absorption, tensile strength, and microstructure evolution of short jute fiber/polylactide composite in a hygrothermal environment	Fabricated using film-stacking hot-pressed method. Aged in a hygrothermal environment	Higher moisture absorption rates for the composites than the neat PLA	Hu et al. [48]
Natural fiber reinforced biodegradable polymer composites	Biodegradation via garbage processing machine	Significant weight reduction of PLA/kenaf composites	Sahari and Sapuan [49]
Advanced materials research	Soil burial test	Higher biodegradation for PLA/kenaf composites than neat PLA	Othman et al. [50]

(Continued)

Table 2.3 (*Continued*)

Title	Manufacturing procedure	Results	Reference
Thermal properties of extruded and injection molded poly(lactic acid)-based cuphea and lesquerella biocomposites	Extruded and then injection molded. Aged for up to 30 days at room temperature (RT)	Extruded materials are more biodegradable than extruded and injection molded materials	Mohamed et al. [51]
Influence of keratin on polylactic acid/chitosan composite properties. Behavior upon accelerated weathering	Melt mixed with Brabender mixer and then compression molded. Exposed to accelerated weathering	Decrease in mechanical properties	Spiridon et al. [36]
Biodegradation of poly(lactic acid)/starch/coir biocomposites under controlled composting conditions	Melt extrusion. Biodegradation study according to ISO 14855 standard	Lower biodegradation rate for compatibilized composites than uncompatibilized composites	Lovino et al. [52]
Beneficial effect of compatibilization on the aging of cellulose reinforced biopolymer blends	Extruded and then compression molded. Aged at 35°C and 80% humidity for 45 days	Cellulosic fillers avert crystallization	Bessadok et al. [6]
Influence of fiber surface treatment on the interfacial property of poly(L-lactic acid)/ramie fabric biocomposites under UV-irradiation hydrothermal aging	Fabricated by hot press molding using the film-stacking procedure. Exposed to UV-irradiation hydrothermal aging	Higher water absorption for composites with treated fiber than that of untreated ones	Chen et al. [53]
Aging of toughened polylactic acid nanocomposites: water absorption, hygrothermal degradation, and soil burial analysis	Extruded and injection molded. Immersed in distilled water and also subjected to soil burial analysis	Lowered the hygrothermal degradation and increased crystallinity for PLA/MMT. Lower biodegradation rate for PLA/MMT than neat PLA	Balakrishnan et al. [22]

Title	Manufacturing procedure	Results	Reference
Processing and properties of PLA/thermoplastic starch/montmorillonite nanocomposites	Melt extrusion. Aged at 258°C and 50% of RH for 300 days	Stable mechanical properties for all composites during 300 days	Arroyo et al. [54]
New polylactide/layered silicate nanocomposites	Extruded and then compression molded. Biodegradation study via industrial composting	Significant increase in biodegradation rate for PLA nanocomposites	Ray et al. [40]
Polylactide/montmorillonite nanocomposites and microcomposites prepared by melt blending: structure and some physical properties	Materials mixed via Brabender mixer and then compression molded	Lower degradation for composites with Na^+ montmorillonite and the organomodified clay	Pluta et al. [55]
Microbial degradation of polyhydroxyalkanoates in tropical soils	Soil burial test	Higher degradation for composites with higher surface area	Boyandin et al. [56]
Comparative study of PHA degradation in natural reservoirs having various types of ecosystems	Biodegradation study in the natural water reservoir	The biodegradability of polymer depends on the environment temperature and the inorganic composition of water	Voinova et al. [57]
Characterizing the effects of ambient aging on the mechanical and physical properties of two commercially available bacterial thermoplastics	Melt extrusion. Aged in an isothermal environment at 15°C	Significant changes in physical and mechanical properties	Srubar et al. [11]
Biodegradability of poly(3-hydroxybutyrate) film grafted with vinyl acetate: effect of grafting and saponification	Enzymatic degradation of PHB-g-VAc powders and films via 0.1 M phosphate buffer	Increased biodegradability for samples saponified in methanol	Wada et al. [58]

(*Continued*)

Table 2.3 (*Continued*)

Title	Manufacturing procedure	Results	Reference
Incorporation of plasticizers in sugarcane-based poly(3-hydroxybutyrate) (PHB): changes in microstructure and properties through aging and annealing	Extruded and then mini injection molded. Aged up to 14 days at RT and RH of 40–50%	Lower properties with aging for pure PHB than for composites with plasticizers	Kurusu et al. [59]
Interfacial improvements in poly(3-hydroxybutyrate)-flax fiber composites with hydrogen bonding additives	A mixture of PHB/fiber in chloroform was cast onto a glass plate to obtain a thin sheet	Higher thermal stability and improved mechanical properties at higher degradation temperatures	Wong et al. [16]
Mechanical and biodegradation performance of short natural fiber polyhydroxybutyrate composites	Extruded and then injection molded. Composted using a rotary aerated composter	Highest biodegradability for PHB/jute composites	Gunning et al. [60]
Effect of compatibilizer content on the mechanical properties of bioplastic composites via hot-melt extrusion	Extruded and then injection molded. Composted using a rotary aerated composter	Higher degradation for compatibilized composites at higher temperatures	Gunning et al. [10]
Assessing biodegradability and mechanical, thermal, and morphological properties of an acrylic acid-modified poly(3-hydroxybutyric acid)/wood flour biocomposite	Biodegradation in soil and enzymatic environments	Higher biodegradation for uncompatibilized composites	Wu [34]
Characterization of polyhydroxybutyrate films and hemp fiber reinforced composites exposed to accelerated weathering	Extruded into polymer tapes. Subjected to accelerated weathering	Decreased mechanical properties	Michel and Billington [61]

Title	Manufacturing procedure	Results	Reference
Thermophysical properties and biodegradation behavior of green composites made from polyhydroxybutyrate and potato peel waste fermentation residue	Extruded and then injection molded. Soil burial test	Decreased mechanical properties and higher degradation rate for composites	Wei et al. [12]
Soil biodegradation of PHBV/peach palm particles biocomposites	Injection molded. Soil burial analysis	Higher biodegradation rate for composites	Batista et al. [35]
The durability of starch-based biodegradable plastics reinforced with Manila hemp fibers	Compression molded. Soil burial analysis	Higher degradation rate for parts of specimens that were under the soil	Ochi [62]
Degradation of naturally occurring polymeric materials in seawater environment	Biodegradation in seawater	Higher degradation for modified cellulose samples than TPS	Rutkowska and Heimowska [63]
Mechanical, moisture absorption, and biodegradation behaviors of bacterial cellulose fiber reinforced starch biocomposites	Via solution impregnation method. Soil burial test and exposed to an environment with RH of 75%	Higher weight loss and mechanical property loss for starch	Wan et al. [9]
Degradation characteristics of SPF/SPS biocomposites	Mixed with mechanical stir and compression molded. Soil burial test and exposed to Xenon arc weathering chamber	Higher biodegradation rate and property loss for SPS compared to SPF/SPS composites	Sahari et al. [64]
The preparation and characterization of silk/gelatin biocomposites	Sandwiched three layers of silk fibers between four films of gelatin. Soil burial test and exposed to simulated weathering for 60 h	Lower degradation rate and less property loss for composites	Shubhra et al. [65]

(Continued)

Table 2.3 (*Continued*)

Title	Manufacturing procedure	Results	Reference
Natural weathering studies of oil palm trunk lumber (OPTL) green polymer composites enhanced with oil palm shell (OPS) nanoparticles	Extruded and impregnated by the vacuum-pressure method. Natural weathering for 6–12 months	Decrease in mechanical properties with increased aging time	Islam et al. [66]
Bio-nanocomposites based on derivatized potato starch and cellulose, preparation and characterization	Solvent evaporation method	Improved mechanical properties without changing the biodegradation rate	Morán et al. [67]

In addition to recyclability, bio-nanocomposites exhibit improvements in mechanical properties, dimensional stability, and solvent or gas resistance and have low density, good transparency, good flow, and better surface properties [41]. Animal-derived protein sources such as casein, whey protein, gelatin, collagen, egg white, and fish myofibrillar are used in degradable bio-nanocomposite materials. Cellulose nanofibers are being used as potential reinforcing materials due to their low cost, biodegradability, and good mechanical properties. Moreover, they have many advantageous properties such as renewability, low weight, low abrasive nature, and good specific properties [42].

The change in the biodegradation rate in nanocomposites is still debatable. In view of present literature studies, it is difficult to make any conclusion about their biodegradation mechanisms. There are no direct investigations conducted to analyze the degradation of starch nanocomposites; consequently, there are no experimental clarifications available regarding the effect of clay on the microbial consumption of starch [43].

2.7 Biodegradation of PLA-Based Composites

PLA is obtained by polymerization of lactic acid. Lactic acid is a sugar fermentation product from corn, sugar beets, sugarcane, and potatoes. The two methods that are currently used in obtaining PLA are polycondensation of lactic acid and via lactic ring opening. PLA is the most widely used biobased and biodegradable polyester. Mainly PLA is used in applications that do not require high performance. Since glass transition temperature (T_g) and melting temperature (T_m) of PLA are between 55–65°C and 150–175°C, respectively, it can be processed by standard methods such as extrusion, injection molding, blow molding, and film-forming operations. PLA has gained tremendous attention as a suitable commodity polymer for replacing petrochemical polymers. However, PLA has some negative aspects such as high cost, high crystallinity and brittleness, and lower molecular weight relative to conventional synthetic plastics [6, 7, 17]. In addition, its biodegradation rate in the soil is very low [13].

Niaounakis et al. [13] aged PLA at different temperatures (20–50°C) at relative humidity (RH) of 80%. The degradation rate was determined by evaluating molecular weight distribution. After 80 days, molecular weight (MW) reduction at 50°C was 82%, while it was less than 30% at 40°C and 20°C. A rapid decrease in properties was discerned for samples that were aged at 50°C for 100 days. After 100 days, 50°C specimens exhibited a 51.4% and 65% decrease in Young's modulus and tensile strength, respectively. At 40°C, Young's modulus decreased by 8% and tensile strength decreased by 30% after 30 days. Only after 130 days, a 20% reduction in properties was observed for 20°C specimens. It was reported that at 50°C, the water acted mainly as a degradation (hydrolysis) agent. Shorter chains, which were generated by PLA degradation, reorganized into crystallites with lower melting points. T_g shifted to lower temperatures with a wide peak for 50°C specimens implying no significant structural changes occurred in the material up to 40°C. Property decrease observed for 40°C specimens was attributed to the faster nucleation and crystallization of the PLA molecules, which was a result of higher molecular mobility due to plasticizing effect of the absorbed water.

Gorrasi et al. [68] compared the hydrolytic behavior of different grades of commercial PLA and they correlated the structural and morphological parameters to the hydrolytic phenomenon. Three commercial grades of PLA produced by NatureWorks LLC, namely, 4060D, 2002D, and 4032D, were used in this work. The average molecular weight of the three grades of PLA is in the range of 190–230 kDa and the polydispersity index is between 1.7 and 1.9. The main difference lies in the D-lactide content of these three grades, which deeply influences the crystallinity of polylactides. It is reported that as the D-lactide percentage increases, the crystallization kinetics rate reduces, and the maximum attainable crystalline content decreases. In this case, 4060D grade is amorphous as it has 12% D-lactide content, 2002D grade is semicrystalline with a maximum crystallinity degree of about 35% having 4% D-lactide, and 4032D is also semicrystalline with a maximum crystallinity degree of about 45% that has 2% D-lactide. Among the four samples, PLA 4060D shows the fastest rate of hydrolytic degradation weight loss and reaches the value of 70% within 30 days as it has the highest enantiomer content. Samples with lower D enantiomer content, irrespective of amorphous or semicrystalline, show very similar and slower rates of weight loss. The 4032D sample, however, shows a slower weight loss revealing a higher resistance to the hydrolytic phenomenon. The optical analysis of the samples also shows that the samples soon become fragile and lose integrity even without any mechanical contact. An increase in pH is shown by the amorphous samples in aqueous media in which the hydrolysis occurred followed by a sharp decrease depending on the D content. Semicrystalline samples show quite a constant value followed by a decreasing trend with a lower rate compared to the amorphous ones. DSC, WAXD, and FTIR results showed that the crystalline content of all samples increased during the hydrolysis test. In a short time, some amorphous regions of all samples change into crystals resulting an increase in the degree of crystallinity due to the erosion of amorphous regions. Over extended time, the crystalline regions undergo hydrolysis.

There are several papers on the degradation mechanisms of PLA, but very little is known about the differences in the degradation mechanism of industrially produced PLA and PLLA (poly-L-lactic acid). Hoglund et al. investigated the difference in the degradation mechanism between industrial-grade PLA and PLLA and explained that industrial PLA is generally a stereocopolymer with some percentage of D-units in the poly(L-lactide) chain with higher molar mass distribution [69]. The stabilizer, nucleating agent, and other additives used in the formulation of industrial PLA also significantly influence the degradation mechanism. Here two types of PLA materials with different D-contents (3051D and 3001D), a 50:50 blend thereof, and synthesized poly(L-lactide) as reference were subjected to hydrolysis at different temperatures and pH values. Though the molar mass of the materials was similar, the molar mass distribution and the D content were different. The mass loss of PLA and PLLA materials was almost the same for the first 49 days, but after that PLA material shows a faster rate of mass loss. The significant differences in the degradation rate of PLA materials can be attributed to the higher degree of crystallinity and lower polydispersity index of PLLA. The induced D-units in the PLA material disturb the crystallization process, and as a result the degree of crystallinity reduces. It is well-known that the degradation of semicrystalline polyesters starts in the amorphous domains and continues in the crystalline domains until all the amorphous domains are degraded. This fact is supported by the evidence that PLLA had a much higher melting temperature and a higher degree of crystallinity prior to degradation as compared to PLA. Those differences were maintained during degradation. But after one day of hydrolytic degradation, the degree of crystallinity for 3051D surprisingly increased from 7% to 26% and 32% in PBS and water, respectively. The aqueous environment and elevated temperature favored the recrystallization process. After day one, the trends in degree of crystallinity are similar for both PLA and PLLA. These findings support the fact that the incorporation of a small number of D-units in the poly(L-lactide) chain augments the hydrolytic degradation process.

The impact of compatibilization was studied by blending sisal fibers (SF) with PLA (PLA/SF) and also with acrylic acid-grafted PLA (PLA-g-AA/SF) [37]. PLA-g-AA/SF (20 wt%) composites showed higher tensile strength (45–50 MPa) compared to PLA/SF (25–30 MPa) due to the greater compatibility and adhesion. The tensile strength of neat PLA was higher than the grafted composites. It was reported that the formation of ester groups as a result of reactions between carboxyl groups of PLA-g-AA and hydroxyl groups in SF was observed, which caused an elevation of tensile strength at 10% SF. After exposing them to soil biodegradation (sample sizes 30 × 30 × 1 mm), PLA-g-AA/SF exhibited a higher water resistance, and its biodegradability rate was lower than that of PLA/SF composites, but higher than that of neat PLA. After 14 weeks, weight loss of AA grafted composites was about 40%, while it was 45–50% for PLA/SF. For neat PLA, it was about 10%. However, both the PLA/SF and the PLA-g-AA/SF composite films were completely degraded. The rate of weight loss of the composites increased with increasing SF content. Similar results were obtained when Wu [33] blended green coconut fiber (GCF) with PLA (PLA/GCF) and also with maleic anhydride (MA) grafted PLA (PLA-g-MA). PLA-g-MA/GCF exhibited higher thermal properties compared to PLA/GCF. After exposing the composites to *Burkholderia cepacia* bacterium (50 × 50 mm sample sizes), it was found that the biodegradability rate of PLA-g-MA/GCF is higher than that of PLA, but lower than that of PLA/GCF, and the rate increased with the addition of GCF. After 21 days, weight loss (%) of PLA, PLA/GCF, and MA-g-PLA/GCF were about 15%, 80%, and 75%, respectively. Improved adhesion between the polymer and the fiber in MA grafted composites was evidenced with increased water resistance in MA-g-PLA/GCF (8%) than in PLA/GCF (9%). Also, poor interaction between PLA and the fiber in PLA/GCF was confirmed by lower crystallinity (25.6 J/g) of PLA/GCF with respect to MA-g-PLA/GCF (32.5 J/g). In both cases, composites with no compatibilizer showed higher weight loss due to hydrophilicity of fiber, which resulted in poor interaction between the matrix and the fiber leading to higher moisture absorption. On the other hand, compatibilized composites exhibited lower weight loss due to improved interfacial adhesion between fiber and polymer matrix.

Silkworm fiber is a natural fiber that has advantageous properties such as biocompatibility and bioresorbable properties that can be utilized in biomedical engineering and surgical operation applications. Ho et al. [45] studied the mechanical properties and biodegradability of 5 wt% silk fiber reinforced PLA composites. It was found that Young's modulus and flexural modulus of the composites increased by 27% and 2%, respectively, while their tensile and flexural strengths decreased (1%). This was attributed to the poor interfacial adhesion between silk fiber and PLA during the mixing process [45]. Specimens were immersed in phosphate buffered saline (PBS) solution to evaluate biodegradability. The composites exhibited higher moisture absorption due to the hydrophilic properties of the silk, resulting in higher biodegradability and reduction in mechanical properties compared to neat PLA. After 4 months, weight increased by 2.4% for the composite, while no change was shown for the neat PLA. Tensile strengths of both PLA and composites decreased to 35–40 MPa and 55–60 MPa, respectively, from 70 MPa. It was reported that water absorption of the samples was due to micromechanical damage in the resin and at the fiber-matrix interface, reduced dimensional stability, and developed internal stresses. Another biodegradability study of PLA/silk fiber composites was conducted by Cheung et al. [46] by using PBS (pH 7.4). At the end of the study, both pure PLA samples and silk/PLA biocomposites did not show considerable differences in pH values and weight loss. After 16 weeks, tensile strengths of both the composite and the neat PLA decreased to 60 MPa from 65 MPa and to 40–43 MPa from 60–62 MPa, respectively, which implies higher property loss in the composites with respect to neat PLA. This was attributed to higher water intake of the composites due to an increase in the area available for hydrolysis with the reinforcement of hydrophilic silk fiber during the 4-month biodegradation period.

PLLA is one of the isomeric forms of PLA. It is homo-crystalline in nature. Chen et al. [53] prepared PLLA blended with ramie fabric biocomposites and studied them under UV-irradiation hydrothermal aging. To enhance the mechanical properties of PLLA, three kinds of ramie fibers were used: (i) untreated ramie fiber – FAB, (ii) ramie fiber treated with permanganate acetone solution – kFAB, and (iii) ramie fiber treated with permanganate acetone solution and silane acetone solution – ksFAB. PLLA/ksFAB exhibited better interfacial adhesion. With respect to PLLA/FAB, flexural and shear strengths of PLLA/kFAB and PLLA/ksFAB biocomposites significantly decreased due to higher water absorption of treated natural fibers than that of untreated ones. Flexural strength decreased from 90 to 80 MPa for PLLA/kFAB, while it dropped to ~65–70 MPa for PLLA/ksFAB. A significant decrease in these properties was exhibited for treated composites after 7 days of aging. For PLLA/FAB, PLLA/kFAB, and PLLA/ksFAB, flexural strength significantly decreased from 90 to 40 MPa, 80 to 35 MPa, and 65–70 to 30–35 MPa, respectively. This sharp decline was attributed to both UV-irradiation and hydrothermal environmental factors. Since there is no pectin, lignin, and other components present in treated fibers, PLLA/ksFAB and PLLA/kFAB were more deteriorated than PLLA/FAB by UV irradiation. The pectin, lignin, and other components have the ability to reduce the destruction caused by UV irradiation [53].

Bessadok et al. [6] studied the effect of compatibilization on the aging of chemically modified cellulose (with 1,4-phenylene diisocyanate) compounded with PLLA-Materbi (PEM) biocomposite. Materbi is a commercially available biodegradable polymer that is made using starches, cellulose, vegetable oils, and their combinations. After aging them at 35°C and 80% humidity for 45 days, the results demonstrated that the addition of cellulosic fillers reduced the degree of crystallinity. For PEM-PLLA (50–50 wt%) composites, crystallinity increased from ~54% to ~79% with aging, while for cellulose/PEM-PLLA and for composites with modified cellulose composites, it was from ~34% to ~38% and from ~50% to ~78%, respectively. Decreased crystallinity for composites with cellulose filler was due to the restriction of polymer chain rearrangement that was caused by fibers during cooling after the extrusion. The increased crystallinity observed for modified cellulose fiber composites was ascribed to the improved compatibilization between the fibers and the matrix. The crystallinity decreased upon aging with decreasing content of PLLA, which confirmed the importance of PLLA during crystallization upon aging [6]. Crystallinity is an important factor in degradation. Higher crystallinity reduces the moisture absorption and microorganism attacks [9]. Consequently, from this study, it can be concluded that if the purpose is to achieve a higher degradation rate, cellulosic filler content plays a major role as it prevents the matrix from crystallizing upon aging.

Montmorillonite (MMT) is used in polymers to improve certain properties, such as barrier properties, flammability resistance, thermal stability, and the rate of biodegradation. PLA/MMT nanocomposites, prepared by Balakrishnan et al. [22], were immersed in distilled water and also subjected to soil burial analysis. When immersed in distilled water, an increase in moisture absorption and hygrothermal stability of PLA were observed. MMT lowered the hygrothermal degradation and increased the crystallinity of PLA. An increase in water absorption was due to the availability of octadecylamine intercalants of MMT for interaction with the water molecules. Moisture absorption was higher at higher temperatures. At room temperature (23°C), PLA exhibited 0.44% of the maximum moisture content (M_m), while it was 0.47% for the composites. At 60°C, M_m increased to 1.4% for PLA, whilst it increased to 1.6% for the composites. T_g of PLA is around 60°C, and the higher vibration and mobility of PLA chains at T_g caused increased water diffusion into the PLA matrix resulting in rapid saturation. Hygrothermal degradation was observed at 60°C, and it was faster at 90°C for both neat PLA and PLA/MMT composites. It was reported that weight loss occurred after saturation point (at 60°C) due to the hydrolysis of PLA. According to soil burial analysis, the biodegradation rate was slightly lower in the PLA/MMT composites (4.45% of weight loss) compared to PLA (4.55% of weight loss). When Arroyo et al. [54] added natural MMT to PLA/thermoplastic starch (TPS) nanocomposites, it increased the tensile modulus by 13% (2.95 to 3.34 GPa) and decreased the elongation at the break by 5–15% of the composites. After aging them at 25°C and 50% RH for 300 days, tensile modulus decreased by 32% (3.34 to 2.95 GPa) for MMT/PLA/TPS, while it exhibited a 15% reduction (2.95 to 2.5 GPa) for PLA/TPS. MMT locates at the blends interface, resulting in weakening the interaction between PLA and TPS phases and thus causing reduced stress transfer from the PLA matrix to the TPS dispersed phase [54].

Aerobic biodegradation of PLA/TPS/Coir natural fibers composite with MA (sample size 10 × 10 × 1 mm) under controlled composting conditions was investigated by Iovino et al. [52]. At the end of the aging period, TPS completely biodegraded due to the microorganisms' attacks on TPS domains, increasing the percentage of CO_2 produced. The CO_2 evolution and degradation of MA grafted composites were lower than those of the uncompatibilized ones. The amount of CO_2 produced for uncompatibilized composites was 101.5 g, while it was 95.3 g for composites with MA. This was attributed to hindered water or microorganism penetration due to improved interface interaction between matrix and fiber [52]. Irregular crystallization temperatures and decreased heat of fusion and melting temperatures (153–147°C) were observed for both biocomposites with respect to neat PLA. It was reported that reduced heat of fusion was due to the disruption of starch and polymer, and decreased melting temperature was a result of the gradual reduction of polymeric chains with increased exposure time. After 90 days, the fibers and the matrix had completely debonded and vanished.

Unidirectional biodegradable composite material fabricated from kenaf fibers and PLA by Sahari and Sapuan [49] showed tensile and flexural strengths of 223 and 254 MPa, respectively. Results of a biodegradability study that was done using a garbage processing machine showed a 38% of weight reduction after four weeks of composting. Another investigation on the biodegradability of PLA/kenaf bast fiber (KBF) was done by Othman et al. [50]. When PLA/KBF films were subjected to a soil burial test, it was found that the decomposition of the composite was faster than pure PLA. According to the SEM morphology of the tensile fracture surface, pores and degradation areas of the composites were larger when KBF content was high. The higher the content of fiber in PLA composite, the higher the micropore surface area of the PLA/KBF biocomposites, which accelerated the disintegration of the biocomposites.

Duigou et al. [47] prepared PLA/flax biocomposites and aged them in natural seawater for 2 years. Flax fibers induced an increase in water uptake of the biocomposites. After two months, water uptake of PLA and PLA/flax was 0.77% and 3.3%, respectively. A linear relationship between water uptake and loss of mechanical properties was observed. After 200 days, Young's modulus and the strength at break decreased by 40%. Unaged PLA/flax biocomposites exhibited a brittle behavior; however, they became more ductile with the increased immersion time. Early damages were observed as the biocomposites underwent aging. The changes were a result of fiber degradation and fiber bundles division, which occurs due to the washing out of soluble components and debonding of fiber bundles. Dried composites were able to regain their portion of mechanical properties, with higher stiffness and lower ductility compared to unaged ones.

A study on the influence of processing techniques on the degradation of PLA/cuphea fiber and PLA/lesquerella fiber composites was conducted by Mohamed et al. [51]. Prepared biocomposites were aged at room temperature for 30 days. It was revealed that the aging process of the biocomposites is affected by extrusion and injection molding. With longer aging time, the percent of crystallinity of PLA in the composites decreased. The enzymatic degradation revealed that extruded composites are more biodegradable than extruded and injection molded composites. Extruded composites exhibited 5–6% weight loss, while it was only 1–2% in extruded and injection molded composites. Unlike extruded composites, extruded and injection molded specimens are covered with PLA and fiber films. Consequently, enzymes have access only from the cut edges of the extruded and injection molded specimens.

Moreover, the weight loss of PLA/lesquerella composites was about 0.2% higher than the PLA/cuphea composites due to the higher content of protein of lesquerella, which made it facile for enzymes to attack.

Spiridon et al. [36] evaluated the influence of keratin fibers on PLA/chitosan composites before and after accelerated weathering. The addition of keratin increased the impact strength and decreased the tensile properties of PLA/chitosan composite. In addition, keratin increased the thermal stability of the composites as keratin acted as a protective blockade on the surface of the material at high temperatures. Upon weathering, impact and tensile strength of PLA decreased from 11 to 4 kJ/m^2 and 59 to 13 MPa, respectively. For PLA/chitosan and PLA/chitosan/keratin composites, tensile strength changed from 36 to 5 MPa and from 50 to 10 MPa, respectively. Impact strength remained constant at 8 kJ/m^2 for composites with keratin, while for PLA/chitosan it changed from 7 to 5 kJ/m^2. The thermal degradation of PLA matrix in the composites was faster than the composites with keratin. After 600 h weathering, for PLA, degradation temperature (T_d) changed from 333°C to 271°C (62°C reduction). When only chitosan was added to PLA, T_d decreased from 318°C to 264°C. Composite with keratin showed a 38°C reduction in T_d (from 317°C to 279°C). According to the results, degradation of PLA matrix under accelerated weathering conditions is higher with chitosan and keratin. After accelerated weathering, the composites showed a rougher surface. It was reported that amorphous PLA crystallized during weathering as polymer chains were shortened due to degradation, which were more prone to crystallize. Composites with no keratin exhibited higher property loss as a result of swelling of material due to water penetration into the hydrophobic matrix and the hydrophilic interface.

Using two lignin types from softwood (LB) and hardwood (LO), Spiridon et al. [32] prepared PLA/LB and PLA/LO composites. The composites exhibited a good adhesion, and the addition of lignin improved the thermal stability and the mechanical properties of PLA. Composites showed a slight increase of Young's modulus due to increased stiffness, a slight decrease in tensile, and a significant increase in impact strength. After 600 h of exposure to accelerated weathering, Young's modulus of the composites exhibited about an 8% increase, while it decreased about 25% for neat PLA. Composites exhibited a 17% decrease in impact strength, whereas it was 60% for neat PLA. Tensile strength reduction for the composites was between 20% and 30%, whilst it was a 75% reduction for pure PLA. As reported, the decrease in properties was due to lessened PLA macromolecule lengths as a result of weathering. However, the increase in Young's modulus was attributed to the recrystallization of material due to weathering conditions.

Jute fibers are commonly used in polymer composites as they are easily available in fiber and fabric forms with good thermal and mechanical properties. Hu et al. [48] aged PLA/jute fiber composites in a hygrothermal environment. The moisture absorption rates of the composites were higher than those of the neat PLA due to the hydrophilic property of the jute fibers owing to the hydroxyl group on the surface of cellulose molecules. It was observed that the moisture absorption process of the composites consisted of three stages: a short and quick moisture absorption stage, a slow stable absorption stage, and an abrupt very rapid moisture absorption stage. The tensile strength of the composites exhibited a 15% decrease after 24 h aging. Fibers and matrix of the PLA/jute composites were fully debonded and PLA turned into scraps by the end of the aging process.

A thermal degradation study of PLA/starch with methylene diphenyl diisocyanate (MDI) was done by Sun et al. [44] by storing them at 25°C with humidity between 30% and 90%. Polyethylene bags were used as moisture barriers for some samples. The thermal decomposition of PLA/starch and PLA/starch/MDI was similar. Thermal endurance of PLA/starch/MDI was greater than that of PLA/starch, but lower than that of neat PLA. As MDI strengthens the interfacial interaction between PLA and starch, the aging rate of PLA/starch/MDI was slower than that of PLA/starch. Thermal-mechanical performances of PLA and its blends were significantly affected by the fluctuated humidity, and the samples stored with polyethylene bags were not affected considerably. Tensile strength of both the PLA blends with and without MDI decreased with longer aging time.

A study conducted by Yatigala [70] included 1–2 wt% maleic anhydride (MA) compatibilized five biobased biodegradable polymers composites with WF (wood fiber). The 60-day soil biodegradation (SB) of these composites (sample size 32 × 32mm) was evaluated against composites without MA under different soil temperatures (30°C and 60°C). The composite samples were prepared with 30 wt% wood fiber and one of the five biopolymers: poly(lactic acid) (PLA), polyhydroxybutyrate (PHB), poly(3-hydroxybutyrate-*co*-3-hydroxyvalerate) (PHBV), Bioflex (BF-PLA blend), and Solanyl (SL-starch based). After 30°C SB, only Solanyl samples showed a significant increase in water absorption (40–50%) due to hydrophilicity of starch-based polymer. After 60°C SB, compatibilized PLA, BF, and PHBV composites exhibited higher or similar water absorption with respect to the composites without MA. Specimens degraded at 60°C showed a significant weight loss percentage as compared to specimens degraded at 30°C (Fig. 2.7).

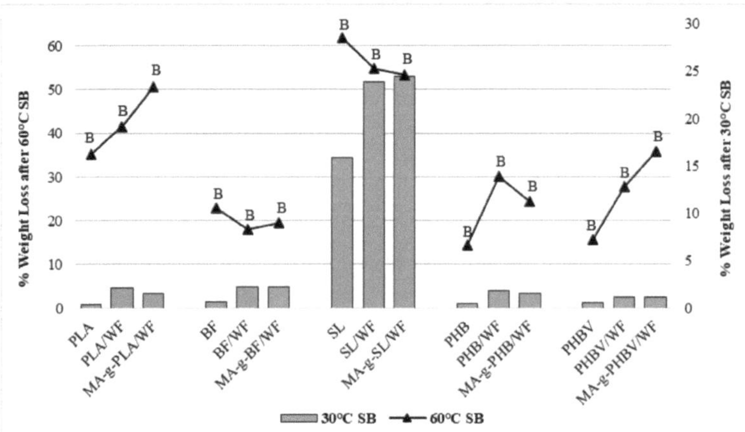

Figure 2.7 Weight loss of samples after 60-day soil burial test at 30°C and 60°C (adapted from [70]).

Cracking and voids were more visible for the specimens buried at 60°C, and they were more fragile as well. For instance, only a few voids were present in neat PLA and its composites after 30°C SB. However, after 60°C SB, both the neat PLA and its composites were extremely fragile with further surface cracks and voids. Compatibilized composites after 60°C SB showed significantly higher degradation than uncompatibilized composites (Table 2.4).

Table 2.4 Images of some samples after 60 days of soil burial test at 30°C and 60°C (adapted from [70])

Composition	Control (before soil burial)	30°C soil burial	60°C soil burial
PLA	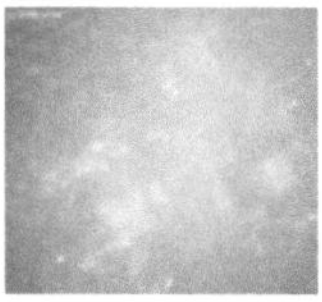	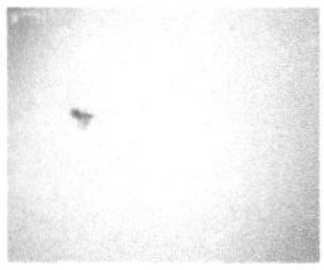	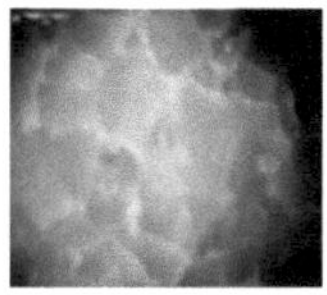
PLA/WF	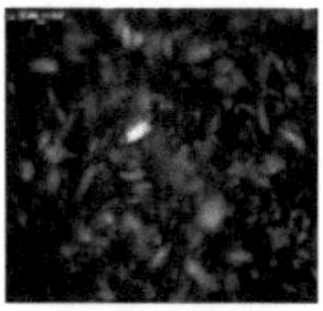	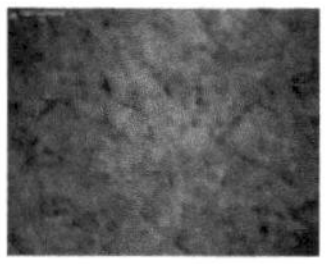	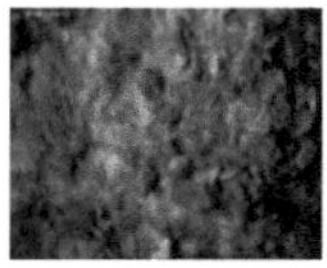
MA-g-PLA/WF	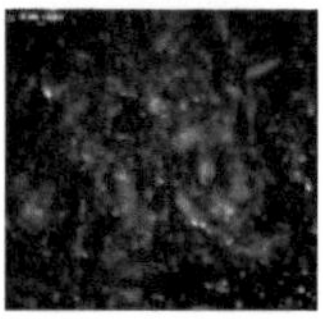	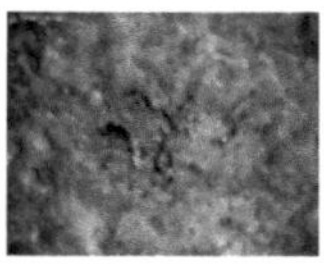	
PHBV		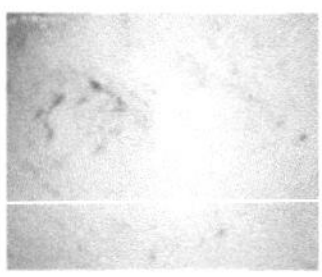	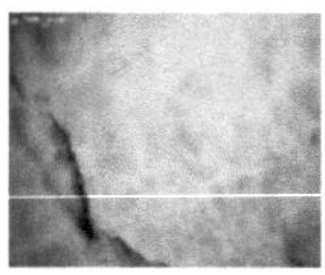
PHBV/WF	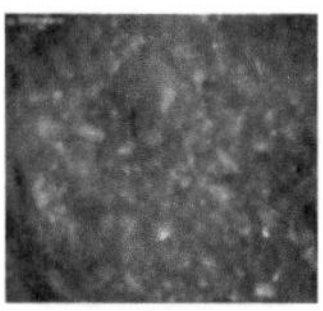		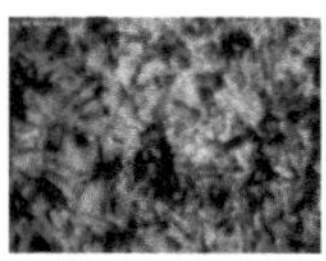
MA-g-PHBV/WF	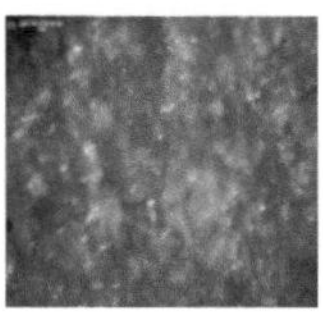	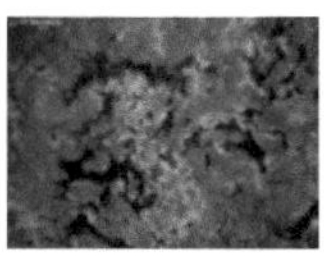	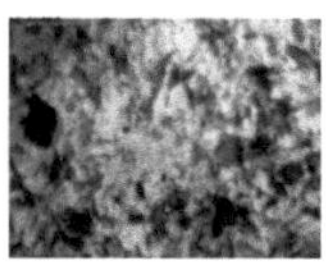

For specimens buried at 30°C, mechanical properties were affected to a lesser degree compared to 60°C SB. In high-temperature degradation environment, the rates of chemical hydrolysis and hydrophilicity of the polymers are high. Microstructural changes and molecular rearrangements occur in the material. When the temperature is higher than the glass transition temperature (T_g) of the polymer, the polymer chains become flexible, increasing water diffusivity to the matrix. The T_g of PLA, BF, and SL polymers are between 55°C and 65°C, and thus water absorption of all these polymer samples was higher at 60°C SB, which resulted in a higher % weight loss compared to 30°C SB. On the other hand, after SB, PHB and PHBV showed a robust structure with very low weight losses at both temperatures. This can be due to their high molecular weights (300,000–400,000 g/mol), which resulted in low solubility of polymers resulting in reduced microbial attacks. The molecular weights of the other three polymers are between 60,000 and 75,000 g/mol. Along with water absorption, biodegradation showed a strong correlation with the molecular weight of these polymers. Composites showed higher degradation than their respective neat polymers due to higher water absorption of hydrophilic WF. Also, degradation was higher in uncompatibilized composites compared to compatibilized composites after 30°C SB, which can be because of reduced water absorption due to improved fiber-matrix adhesion. However, this was reversed after 60°C SB. At higher temperatures, covalent ester carbonyl bonds between the polymer and the fiber hydrolyze. This allows unconstrained fibers and matrix to absorb more water. When fiber and matrix are debonded where fibers are enclosed well with the polymer due to compatibilization, biodegradation can be higher due to the higher surface area available for biodegradation. This is visible in MA-g-PLA/WF samples after 60°C SB (Table 2.4).

2.8 Biodegradation of PHB and PHBV-Based Composites

PHB and PHBV are important examples of PHA bioplastics. PHAs are produced in nature by bacterial fermentation of sugar or lipids. PHB is accumulated as a reserve of carbon and energy by a number of bacteria. PHBV copolymer is produced by incorporating 3-hydroxyvalerate (3HV) units into PHB segments [71]. PHB is degraded by numerous microorganisms in various ecosystems: *Alternaria* sp., *Penicillium simplicissimum*, *P. funiculosum*, *P. notatum*, and *Eupenicillium* sp. [28]. During degradation, PHA polymer is considerably excessive in size to be conveyed directly through the bacterial cell wall, so it should be permuted into corresponding hydroxyl acid monomers. In order to do that, bacteria evolve extracellular hydrolases [72]. These monomers are soluble in water and small enough to diffuse through the cell wall. Under aerobic conditions, they produce CO_2 and water by being metabolized by β-oxidation and tricarboxylic acid cycle, and under anaerobic conditions, methane is also produced [15]. The product of PHB hydrolysis is R-3-hydroxybutyric acid [73], and for PHBV it is both 3-hydroxybutyrate and 3-hydroxyvalerate. Biodegradation of PHAs depends on many factors such as microbial activity, moisture, temperature, pH of the environment, the exposed surface area, molecular weight polymer composition, the nature of the monomer unit, and crystallinity [8].

Boyandin et al. [56] investigated the biodegradation of PHA films and pellets by soil microorganisms in tropical Vietnamese soils. Results revealed that the degradation of PHA depends on polymer chemical composition, specimen shape, and microbial characteristics. The average mass loss rates were 0.04–0.33% and 0.02–0.18% per day for films and compact pellets, respectively. Higher degradation of PHA film was attributed to the availability of a larger surface area of films (provides a larger polymer-soil interface area), which facilitated the better attachment of microorganisms and faster biofilm formation. A 20–60% decrease in the polymer molecular mass and a 2–3% increase in the degree of crystallinity were observed during the degradation of both PHA films and pellets, suggesting preferential degradation in the amorphous phase. When Voinova et al. [57] studied the kinetics of PHA degradation in the natural environment in water reservoirs, it was revealed that biodegradation of polymer depends on the environment temperature and inorganic composition of water. After 49 days at 19°C water temperature, polymers degraded almost completely, having a residual weight of 2.3–5.2% of the initial value. The half-life of PHA at 19°C was 32 days while it was 55 days at 5°C.

Srubar et al. [11] aged PHB and PHBV in an isothermal environment at 15°C, which is moderately above their T_g. A significant increase in tensile modulus of elasticity and a decrease in ultimate strain were observed. Over the testing period (168 days), PHB and PHBV samples had a maximum increase in modulus of 166% and 178% and crystallinity of approximately 41% and 58%, respectively. After 168 days, the elongation-to-break of PHB and PHBV samples decreased by 64% and 72% and the ultimate strengths of the samples decreased by 28% and 8.9%, respectively. Melt extrusion caused a minor reduction in molecular weight, and during the aging process, the amorphous regions of PHB and PHBV underwent glassy aging in the rigid amorphous fraction, resulting in a significant embrittlement. This transition eliminated the excess free volume, toughening the amorphous material. This could be the reason for the increase in tensile modulus and decrease in the strain at break [11].

PHB with flax fiber composites was fabricated by Wong et al. [16]. Flax fibers were treated with 4,4'-thiodiphenol (TDP) as a hydrogen bonding additive. The hydrogen bonding between the carboxyl group of PHB and the OH group of the fibers exhibited a positive effect on dynamic flexural properties. Also, it showed improvements in storage modulus, composite stiffness, T_g, and thermal stability of the composites at higher degradation temperatures. When Gunning et al. [60] prepared three different biocomposites by blending hemp, jute, and lyocell fibers with PHB, melt flow index, impact, and tensile strength of the blends decreased, and PHB/hemp had the smallest drop. Figure 2.8 shows the rate of composting (using a rotary aerated composter) of these composites containing 30 wt% of fiber. PHB/jute composites exhibited the highest level of degradation due to increased water absorption and greater dispersion of fibers. Higher water absorption causes the debonding of the materials. When the material debonds, increased fiber dispersion increases the availability of surface area for microbial attack, resulting in a higher degradation rate. Furthermore, the drop in week 3 or 4 was due to the increased temperature (60°C), which caused the destruction of some microorganisms. However, as the new bacteria grew, the degradation rate increased. After 12 weeks, PHB/jute composites showed a 34% increase in biodegradation compared to neat PHB. When Gunning et al. [10] used PHB-g-MA with the same fiber types, composites showed improved mechanical properties, but lower degradation rates. However, when the temperature increased to 60°C during composting, the rate of degradation of the composites with MA was higher than those without. After 12 weeks of composting, PHB, PHB/jute, and MA-g-PHB/jute showed a weight loss of 35%, 50%, and 80%, respectively. As the temperature increased, hydrolysis of the ester bond took place between MA and fiber. Since the MA increased fiber dispersion in the matrix compared to uncompatibilized ones, when the fiber debonded from the matrix at higher temperatures, it left a higher surface area available for microbial attack accelerating the degradation rate.

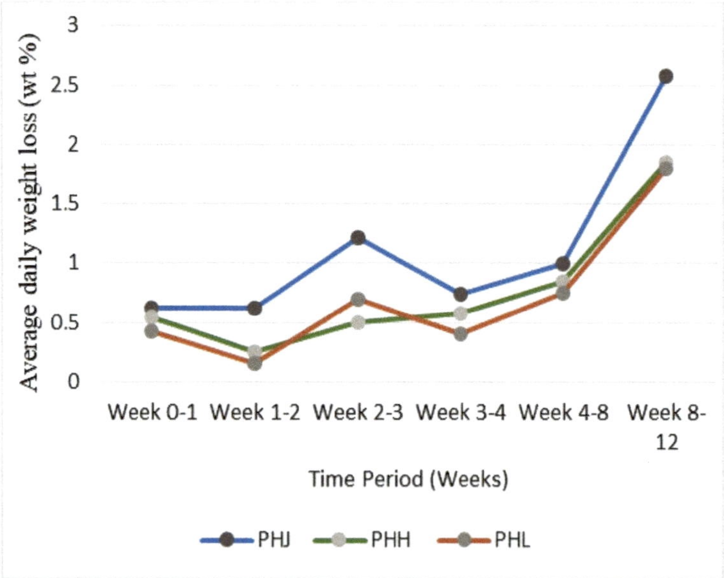

Figure 2.8 Weight loss of PHB/jute, PHB/hemp, and PHB/Lyocell fiber composites, containing 30 wt% of fiber.

Wu [34] prepared acrylic acid (AA) grafted PHB/wood flour (WF) composites and subjected them to biodegradation studies in soil and enzymatic environments. PHB/WF had the highest degradation rate in both environments. After 12 weeks, the residual weight percentage of PHB/WF and AA-g-PHB/WF in soil was about 72% and ~78%, respectively. As the wood flour content increased, the degradation rate of the composites increased due to hindrance in forming polymer chain arrangements. The residual weight of 50 wt% fiber composites was about 20% lower than composites with 30 wt% fiber. Water resistance was higher in PHB-g-AA/WF than in PHB/WF composites, but lower than in neat PHB. Weight gain % of PHB was only about 1.5%, while it was about 7% and 8% for AA-g-PHB/WF and PHB/WF, respectively. The hydrophilic nature of

WF and hydrophobic PHB caused poor adhesion, which increased the water absorption resulting in an increased rate of degradation. Lower weight loss of AA-g-PHB/WF (with respect to PHB/WF) was attributed to increased water resistance due to the ester carbonyl functional group of that blend.

PHB and PHB/hemp fiber composites were subjected to accelerated weathering by Michel and Billington [61] under two weathering conditions: with cyclic elevated relative humidity (P1 – UV, heat, moisture) and one without (P2 – UV, heat). The length of the P1 weathering procedure was 1998 h, while it was 384 h for P2. With weathering exposure, both the polymer and composites exhibited mass loss and increased fading. The fading was a result of UV absorption by chromophores of hemp fibers and PHB, and also the increased diffuse reflectance from polymer crazing. Mass loss of 8% for P1 PHB specimens and 4.2% for P2 PHB specimens were observed. This was attributed to surface erosion and reduction in molecular weight occurred due to UV photooxidative and cyclic hygrothermal degradation mechanism. P1 composites first exhibited an increase in mass, and then the mass decreased to its original weight by the end of weathering. The increase in mass was due to the hygroscopicity of the hemp fibers. The maximum mass loss for P2 composites was 1.4%. Mass loss for both composites was attributed to weakened adhesion between the matrix and the fiber due to different coefficients of thermal expansion of the polymer matrix and the fiber. In both weathering exposures, the ultimate tensile strength of both specimens decreased. For PHB, 30% and 42% of decrement was exhibited in both P1 and P2 conditions, respectively. For the composites, the decrement was 47% and 25% for P1 and P2, respectively. For PHB, this was a result of molecular weight drop due to polymer chain scission from UV photooxidation and the hydrolytic attack that occurred because of the high moisture environment. As for the PHB/hemp composites, it was attributed to fiber swelling and brittleness of the biopolymer matrix occurred due to photooxidation and hydrolysis.

Batista et al. [35] subjected PHBV/peach palm particles (PPp) biocomposites to soil burial analysis (sample size 30 × 20 × 3 mm). When PPp was incorporated into PHBV, a reduction in the maximum strength and the elongation at break, and an improvement in Young's modulus of the biocomposites were observed. With 10% (w/w) PPp in the PHBV matrix, the tensile strength of the composites exhibited a 35% reduction. When the PPp content increased to 25% (w/w), no further reduction in tensile strength was observed. Soil biodegradation was evaluated by analyzing visual appearance. Biodegradation increased with increasing content of PPp. According to SEM, the distance between the PPp and the matrix in the composites amplified as the PPp content was increased. The higher the PPp content in the composites, the higher the moisture absorption and microbial attack due to poor adhesion between the PPp and the polymer matrix, which results in higher biodegradation.

Wei et al. [12] fabricated PHB and potato peel waste fermentation residue (PPW-FR) composites and subjected them to soil burial analysis (sample size 24 × 2 mm). Biodegradation of the specimens was determined by visual appearance. After 8 months, composites with higher filler content (50 wt%) exhibited absolute debonding, while lower (<50 wt%) fiber content composites exhibited only partial degradation. From microscopic analysis, it was observed that the first component that was degraded was PPW-FR, resulting in a higher amount of PHB and pores on the surface of the specimens. In the case of PHB, amorphous regions were degraded first followed by crystalline regions.

Kurusu et al. [59] incorporated three types of plasticizers (10 wt%), namely, tri(ethylene glycol) bis(2-ethylhexanoate) (TEG-EH), triethyl citrate (TEC), and tributyl citrate (TBC), to sugarcane-based PHB to increase toughness. Annealing treatment was used to increase the toughness and avoid the aging of some PHB-based materials. The composites exhibited an increase in impact strength, but poor mechanical performance. Deterioration of properties with aging was more significant for pure PHB than for composites with plasticizers. PHB/TEG-EH had faintly better mechanical performance compared to other compositions. It was stated that this could be attributed to the processing technique (twin-screw extrusion), which was used for plasticizer incorporation. Mechanical properties of the samples decreased more after annealing as the migration of plasticizers from the interlamellar amorphous regions occurred.

2.9 Biodegradation of Polymers Extracted from Biomass

There are many other degradation studies conducted on biobased composites manufactured from natural polymers such as gelatin, oil palm shell, different kinds of starches, and many natural fibers such as hemp, cellulose, silk, and sugar palm fibers.

Manila hemp fiber reinforced starch-based plastics were studied by Ochi [62] in natural soil (only half-buried in soil) and in compost soil (completely buried). The biodegradability and mechanical properties of the parts of specimens (sample size 10 × 200 × 1 mm) that were under soil changed significantly, while it was insignificant in the atmosphere. Tensile strength of composted and natural soil buried composites decreased 80% after 20 days and 90 days, respectively. Weight loss of composted specimens was higher than the half-buried composites. Composted specimens lost 30% of weight at 90°C in 30 days, while the natural soil buried specimens had the same weight loss in 180 days. When the specimen interaction with microbial is higher, significant loss of weight and tensile strength can be observed. Wan et al. [9] prepared bacterial cellulose nanofibers (BC) reinforced starch composites; they showed a higher tensile strength and Young's modulus compared to unreinforced starch. The reduction of tensile strength after 21 days in soil for starch was 79%, while it was 69% for the composite. For starch, the average degradation rate in soil was about 1% per day, though it was 0.9% per day for the composite. This was attributed to the higher crystallinity of BC, which resulted in higher resistance to moisture and microorganism attacks.

Sugar palm fiber is used as a reinforcement material due to its high durability and resistance to seawater. Sahari et al. [64] prepared plasticized (with glycerol) sugar palm fiber (SPF) reinforced sugar palm starch (SPS) composites. They were exposed to a Xenon arc weathering chamber and subjected to soil burial experiments. After 72 h of soil burial, neat SPS lost 63.58% of its weight, while it was 56.73% for SPF/SPS composites. Hence, the biodegradation rate of SPS was higher compared to the SPF/SPS biocomposites. This was attributed to the higher water intake of SPS, which made it more prone to microorganism attacks. At the end of 72 h of the weathering,

SPS and SPF/SPS biocomposites lost 78.09% and 53.67% of tensile strength, respectively. This was due to the degradation of both SPS and SPF caused by UV radiation [64].

Shubhra et al. [65] fabricated silk fiber/gelatin biocomposites. Composites containing 30% of silk exhibited a notable improvement in mechanical properties; however, the water uptake was much lower (80% in 20 min) than that of gelatin (350% in 20 min). After exposing them to a simulated weathering tester for 60 h, it was observed that the tensile strength loss of the composites increased with decreased fiber content. Tensile strength of 30 wt% silk composites decreased by 70% and for 10 wt% silk composites by 89%. During soil biodegradation, it was observed that with increasing content of fiber, weight loss of the composites decreased. After 24 h soil burial, the tensile strength of gelatin/silk (30 wt%) decreased to 20 MPa and lost 51% weight. For composites with 10% fibers, weight loss was 60%. Increased content of silk caused to increase the bonding with gelatin, reducing the void spaces, which resulted in higher tensile properties. It was explained that during the soil burial test, the structure of the composites was loosened due to the formation of cross-links between water molecules and fibroin chains. Fiber strength properties decreased due to primary bonds scission in the backbone of macromolecules and/or secondary bonds breakage due to reactive species such as water.

Islam et al. [66] exposed oil palm trunk lumber (OPTL)/ oil palm shell (OPS) composites to natural weathering for 6–12 months in Indonesia. Grounded OPS fibers were mixed with phenol formaldehyde (PF) before impregnating to OPTL. According to the results, both the exposure time and the concentration of PF/OPS had a significant effect on the properties of OPTL. The mechanical, physical, and morphological properties of the samples decreased with increased exposure time. The average weight losses were between 27% and 36% for all composites (1–10% PF-OPS) after 12-month exposure. The weight loss for neat OPTL was about 42%. The mechanical properties of all samples deteriorated. After 6–12 months of weathering exposure, tensile strength, flexural strength, and impact strength of composites with 5% PF-OPS decreased from 4.8% to 11.1%, 2.4% to 4.4%, and 8.0% to 13.3%, respectively. These changes were relatively higher for neat OPTL. The reason for having better properties for the composite than the neat OPTL is that PF-

OPS nanoparticles filled the cell lumen forming a rigid cross-linked polymer, which resulted in improved strength properties. With the increasing exposure time, loss of properties and degradation of the polymer occur due to chemical bond scission reactions in macromolecules.

Rutkowska and Heimowska [63] aged TPS/modified cellulose biocomposites in Baltic Sea at Nordic Wharf of Gdynia harbor. Samples showed a clear erosion of the surface and weight loss. They reported the occurrence of enzymatic hydrolysis. Modified cellulose was more affected by microbial attacks in seawater than TPS. The film form of starch/modified cellulose composites showed more distinct weight changes than its sheet form due to large surface development at the phase boundary.

Wheat gluten-based bioplastics plasticized with glycerol were subjected to biodegradation in liquid medium (modified Sturm test) and in farmland soil [74]. The mineralization of the sample in a liquid medium was determined by measuring the amount of CO_2 formed, while in the soil it was measured with weight loss. Complete degradation of all gluten materials was observed after 36 days in aerobic fermentation and after 50 days in soil. There were no significant differences between the samples. The possibility of toxic metabolites formation during degradation was examined by carrying out microbial inhibition, and it showed no signs of toxic effects on the modified gluten or on its metabolites.

2.10 Conclusion

Biodegradable biobased polymers have gained much interest recently due to their low impact on the environment. There are several articles that highlight the importance of utilizing biobased polymers and their composites from renewable resources instead of petroleum-based polymers. However, along with the desirable properties of biobased polymers, they have disadvantageous physical properties in terms of strength, dimensional stability, and difficulty in processing. Even though biopolymers have numerous disadvantages, these can be blended with other natural polymers to create composites with unique properties that do not exist naturally. Nonetheless, biodegradation of plastics requires certain conditions; biodegradable polymers do not necessarily degrade in any environment or under any conditions. During biodegradation, due to microorganisms' attack and disintegration activity on polymeric materials, the polymer's physical and chemical properties and molecular mass deteriorate forming CO_2, H_2O, CH_4, and other low molecular weight products as end products.

When fibers are added to polymers such as PLA, PHB, and PHBV, an increase in degradation rate can be observed with increasing content of fiber due to higher water absorption. Consequently, fibers lower the composite's mechanical properties such as tensile strength, elongation at break, and impact strength. The main reason for these poor properties of PLA, PHB, and PHBV based fiber biocomposites is the poor adhesion between the polymer matrix and the fiber due to the hydrophobic nature of the polymer matrix and the hydrophilic nature of the fibers. Nonetheless, a greater adhesion can be achieved by introducing reactive functional groups such as compatibilizers to the composites. Even though compatibilizers improve the properties of the composites, they reduce the degradation rate due to higher water resistance. However, the degradation rates of these compatibilized composites are still higher than those of the neat polymers. Nevertheless, when aged, both compatibilized and uncompatibilized biopolymer composites have shown a complete debonding of fibers and matrix and a complete degradation after a specified period. When the composites are exposed to weathering, they exhibit mass loss, fading, and a decrease in strength properties.

Abbreviations

AA	Acrylic acid
DMA	Dynamic mechanical analyzer
DSC	Differential scanning calorimetry
MA	Maleic anhydride
MWD	Molecular weight distribution
PHA	Polyhydroxyalkanoate
PHB	Poly(hydroxybutyrate)
PHBV	Poly(3-hydroxybutyrate-*co*-3-hydroxyvalerate)
PLA	Polylactic acid
RH	Relative humidity
RT	Room temperature

SEM	Scanning electron microscopy
T_{cc}	Cold crystallization temperature
T_g	Glass transition temperature
T_m	Melting temperature
TGA	Thermogravimetric analysis
TPS	Thermoplastic starch

References

1. Madbouly, S., Schrader, J., Srinivasan, G., Haubric, K., et al., *2012 Bioplastic Container Cropping Systems Conference*, Iowa State University 2012.
2. Rathje, W. L. and Murphy, C., 2001. *Rubbish!: The Archaeology of Garbage*. The Plastics Industry Trade Association, Harper Collins, University of Arizona Press.
3. Bronzino, J. D., *Biomedical Engineering Handbook*, CRC Press 1999.
4. U.S Congress, O. o. T. A., Washington, DC. Proceedings of Congress and General Congressional Publications Volume 139, Part 13 (August 2, 1993 to August 5, 1993).
5. Suryanegara, L., Nakagaito, A. N., Yano, H., The effect of crystallization of PLA on the thermal and mechanical properties of microfibrillated cellulose-reinforced PLA composites. *Composites Science and Technology* 2009, *69*, 1187–1192.
6. Bessadok, A., Belgacem, M. N., Dufresne, A., Bras, J., Beneficial effect of compatibilization on the aging of cellulose-reinforced biopolymer blends. *Macromolecular Materials and Engineering* 2010, *295*, 774–781.
7. Petinakis, E., Simon, G., Dean, K., Yu, L., *Natural Fibre Bio-Composites Incorporating Poly (Lactic Acid)*, INTECH Open Access Publisher 2013.
8. Bugnicourt, E., Cinelli, P., Lazzeri, A., Alvarez, V., Polyhydroxyalkanoate (PHA): review of synthesis, characteristics, processing and potential applications in packaging. *Express Polym. Lett* 2014, *8*, 791–808.
9. Wan, Y., Luo, H., He, F., Liang, H., et al., Mechanical, moisture absorption, and biodegradation behaviours of bacterial cellulose fibre-reinforced starch biocomposites. *Composites Science and Technology* 2009, *69*, 1212–1217.
10. Gunning, M. A., Geever, L. M., Killion, J. A., Lyons, J. G., Higginbotham, C. L., Effect of compatibilizer content on the mechanical properties of bioplastic composites via hot melt extrusion. *Polymer-Plastics Technology and Engineering* 2014, *53*, 1223–1235.

11. Srubar, W., Wright, Z., Tsui, A., Michel, A., et al., Characterizing the effects of ambient aging on the mechanical and physical properties of two commercially available bacterial thermoplastics. *Polymer Degradation and Stability* 2012, *97*, 1922–1929.
12. Wei, L., Liang, S., McDonald, A. G., Thermophysical properties and biodegradation behavior of green composites made from polyhydroxybutyrate and potato peel waste fermentation residue. *Industrial Crops and Products* 2015, *69*, 91–103.
13. Niaounakis, M., Kontou, E., Xanthis, M., Effects of aging on the thermomechanical properties of poly (lactic acid). *Journal of Applied Polymer Science* 2011, *119*, 472–481.
14. Azwa, Z., Yousif, B., Manalo, A., Karunasena, W., A review on the degradability of polymeric composites based on natural fibres. *Materials & Design* 2013, *47*, 424–442.
15. Shah, A. A., Hasan, F., Hameed, A., Ahmed, S., Biological degradation of plastics: a comprehensive review. *Biotechnology Advances* 2008, *26*, 246–265.
16. Wong, S., Shanks, R., Hodzic, A., Interfacial improvements in poly (3-hydroxybutyrate)-flax fibre composites with hydrogen bonding additives. *Composites Science and Technology* 2004, *64*, 1321–1330.
17. Mukherjee, T., Kao, N., PLA based biopolymer reinforced with natural fibre: a review. *Journal of Polymers and the Environment* 2011, *19*, 714–725.
18. Pilla, S., *Handbook of Bioplastics and Biocomposites Engineering Applications*, Scrivener Publishing, Salem, MA.
19. Vroman, I., Tighzert, L., Biodegradable polymers. *Materials* 2009, *2*, 307–344.
20. Schnabel, W., *Polymer Degradation*, Macmillan Publishing 1981.
21. Hamid, S. H. *Handbook of Polymer Degradation*, CRC Press, 2000.
22. Balakrishnan, H., Hassan, A., Imran, M., Wahit, M. U., Aging of toughened polylactic acid nanocomposites: water absorption, hygrothermal degradation and soil burial analysis. *Journal of Polymers and the Environment* 2011, *19*, 863–875.
23. Proikakis, C., Mamouzelos, N., Tarantili, P., Andreopoulos, A., Swelling and hydrolytic degradation of poly (D, L-lactic acid) in aqueous solutions. *Polymer Degradation and Stability* 2006, *91*, 614–619.
24. Kister, G., Cassanas, G., Bergounhon, M., Hoarau, D., Vert, M., Structural characterization and hydrolytic degradation of solid copolymers of d, l-lactide-co-ε-caprolactone by Raman spectroscopy. *Polymer* 2000, *41*, 925–932.

25. Lucas, N., Bienaime, C., Belloy, C., Queneudec, M., et al., Polymer biodegradation: mechanisms and estimation techniques – A review. *Chemosphere* 2008, *73*, 429–442.

26. Jakubowicz, I., Yarahmadi, N., Petersen, H., Evaluation of the rate of abiotic degradation of biodegradable polyethylene in various environments. *Polymer Degradation and Stability* 2006, *91*, 1556–1562.

27. Leja, K., Lewandowicz, G., Polymer biodegradation and biodegradable polymers—a review. *Polish Journal of Environmental Studies* 2010, *19*, 255–266.

28. Kim, M.-N., Lee, A.-R., Yoon, J.-S., Chin, I.-J., Biodegradation of poly (3-hydroxybutyrate), Sky-Green® and Mater-Bi® by fungi isolated from soils. *European Polymer Journal* 2000, *36*, 1677–1685.

29. Müller, R. J., Biodegradability of polymers: regulations and methods for testing. *Biopolymers Online* 2005.

30. Biodegradable Products Institute, Inc, https://products.bpiworld.org/

31. Nayaran, R., Pettigrew, C., ASTM Standards Help Define and Grow a New Biodegradable Plastic Industry. *ASTM Standardization News (December, 1999).*

32. Spiridon, I., Leluk, K., Resmerita, A. M., Darie, R. N., Evaluation of PLA–lignin bioplastics properties before and after accelerated weathering. *Composites Part B: Engineering* 2015, *69*, 342–349.

33. Wu, C.-S., Renewable resource-based composites of recycled natural fibers and maleated polylactide bioplastic: characterization and biodegradability. *Polymer Degradation and Stability* 2009, *94*, 1076–1084.

34. Wu, C. S., Assessing biodegradability and mechanical, thermal, and morphological properties of an acrylic acid-modified poly (3-hydroxybutyric acid)/wood flours biocomposite. *Journal of Applied Polymer Science* 2006, *102*, 3565–3574.

35. Batista, K., Silva, D., Coelho, L., Pezzin, S., Pezzin, A., Soil biodegradation of PHBV/peach palm particles biocomposites. *Journal of Polymers and the Environment* 2010, *18*, 346–354.

36. Spiridon, I., Paduraru, O. M., Zaltariov, M. F., Darie, R. N., Influence of keratin on polylactic acid/chitosan composite properties. Behavior upon accelerated weathering. *Industrial & Engineering Chemistry Research* 2013, *52*, 9822–9833.

37. Wu, C. S., Preparation, characterization, and biodegradability of renewable resource-based composites from recycled polylactide

bioplastic and sisal fibers. *Journal of Applied Polymer Science* 2012, *123*, 347–355.

38. Joshi, S., Can nanotechnology improve the sustainability of biobased products? *Journal of Industrial Ecology* 2008, *12*, 474–489.
39. Maiti, P., Batt, C., Renewable plastics: synthesis and properties of PHB nanocomposites. *Polymeric Materials Science and Engineering* 2003, *88*, 58–59.
40. Ray, S. S., Yamada, K., Okamoto, M., Fujimoto, Y., et al., New polylactide/layered silicate nanocomposites. 5. Designing of materials with desired properties. *Polymer* 2003, *44*, 6633–6646.
41. Zhao, R., Torley, P., Halley, P. J., Emerging biodegradable materials: starch-and protein-based bio-nanocomposites. *Journal of Materials Science* 2008, *43*, 3058–3071.
42. Kalia, S., Dufresne, A., Cherian, B. M., Kaith, B., et al., Cellulose-based bio-and nanocomposites: a review. *International Journal of Polymer Science* 2011, *2011*.
43. Pandey, J. K., Reddy, K. R., Kumar, A. P., Singh, R., An overview on the degradability of polymer nanocomposites. *Polymer Degradation and Stability* 2005, *88*, 234–250.
44. Acioli-Moura, R., Sun, X. S., Thermal degradation and physical aging of poly (lactic acid) and its blends with starch. *Polymer Engineering & Science* 2008, *48*, 829–836.
45. Ho, M.-p., Lau, K.-t., Wang, H., Bhattacharyya, D., Characteristics of a silk fibre reinforced biodegradable plastic. *Composites Part B: Engineering* 2011, *42*, 117–122.
46. Cheung, H.-Y., Lau, K.-T., Pow, Y.-F., Zhao, Y.-Q., Hui, D., Biodegradation of a silkworm silk/PLA composite. *Composites Part B: Engineering* 2010, *41*, 223–228.
47. Le Duigou, A., Bourmaud, A., Davies, P., Baley, C., Long term immersion in natural seawater of Flax/PLA biocomposite. *Ocean Engineering* 2014, *90*, 140–148.
48. Hu, R.-H., Sun, M.-y., Lim, J.-K., Moisture absorption, tensile strength and microstructure evolution of short jute fiber/polylactide composite in hygrothermal environment. *Materials & Design* 2010, *31*, 3167–3173.
49. Sahari, J., Sapuan, S., Natural fibre reinforced biodegradable polymer composites. *Reviews on Advanced Materials Science* 2011, *30*, 166–174.
50. Othman, M., Ibrahim, N. A., Ruzaidi, C. M., Zakaria, N., and Halim, Z., 2012. Biodegradability analysis of KBF reinforced poly(lactic acid) biocomposites. *Advanced Materials Research*, *576*, 434–437.

51. Mohamed, A., Finkenstadt, V., Rayas-Duarte, P., Debra, E., Gordon, S. H., Thermal properties of extruded and injection-molded poly (lactic acid)-based cuphea and lesquerella bio-composites. *Journal of Applied Polymer Science* 2009, *111*, 114–124.
52. Lovino, R., Zullo, R., Rao, M., Cassar, L., Gianfreda, L., Biodegradation of poly (lactic acid)/starch/coir biocomposites under controlled composting conditions. *Polymer Degradation and Stability* 2008, *93*, 147–157.
53. Chen, D., Li, J., Ren, J., Influence of fiber surface-treatment on interfacial property of poly (l-lactic acid)/ramie fabric biocomposites under UV-irradiation hydrothermal aging. *Materials Chemistry and Physics* 2011, *126*, 524–531.
54. Arroyo, O., Huneault, M., Favis, B., Bureau, M., Processing and properties of PLA/thermoplastic starch/montmorillonite nanocomposites. *Polymer Composites* 2010, *31*, 114–127.
55. Pluta, M., Galeski, A., Alexandre, M., Paul, M. A., Dubois, P., Polylactide/montmorillonite nanocomposites and microcomposites prepared by melt blending: structure and some physical properties. *Journal of Applied Polymer Science* 2002, *86*, 1497–1506.
56. Boyandin, A. N., Prudnikova, S. V., Karpov, V. A., Ivonin, V. N., et al., Microbial degradation of polyhydroxyalkanoates in tropical soils. *International Biodeterioration & Biodegradation* 2013, *83*, 77–84.
57. Voinova, O., Gladyshev, M., Volova, T. G., *Macromolecular Symposia*, Wiley Online Library 2008, pp. 34–37.
58. Wada, Y., Seko, N., Nagasawa, N., Tamada, M., et al., Biodegradability of poly (3-hydroxybutyrate) film grafted with vinyl acetate: effect of grafting and saponification. *Radiation Physics and Chemistry* 2007, *76*, 1075–1083.
59. Kurusu, R. S., Siliki, C. A., David, É., Demarquette, N. R., et al., Incorporation of plasticizers in sugarcane-based poly (3-hydroxybutyrate) (PHB): changes in microstructure and properties through ageing and annealing. *Industrial Crops and Products* 2015, *72*, 166–174.
60. Gunning, M. A., Geever, L. M., Killion, J. A., Lyons, J. G., Higginbotham, C. L., Mechanical and biodegradation performance of short natural fibre polyhydroxybutyrate composites. *Polymer Testing* 2013, *32*, 1603–1611.
61. Michel, A., Billington, S., Characterization of poly-hydroxybutyrate films and hemp fiber reinforced composites exposed to accelerated weathering. *Polymer Degradation and Stability* 2012, *97*, 870–878.

62. Ochi, S., Durability of starch based biodegradable plastics reinforced with Manila hemp fibers. *Materials* 2011, *4*, 457–468.
63. Rutkowska, M., Heimowska, A., Degradation of naturally occurring polymeric materials in sea water environment. *Polimery* 2008, *53*, 854–864.
64. Sahari, J., Sapuan, S. M., Zainudin, E. S., Maleque, M. A., Degradation characteristics of SPF/SPS biocomposites. *Fibres & Textiles in Eastern Europe* 2014, *22*, 96–98.
65. Shubhra, Q. T., Alam, A., Khan, M., Saha, M., et al., The preparation and characterization of silk/gelatin biocomposites. *Polymer-Plastics Technology and Engineering* 2010, *49*, 983–990.
66. Islam, M. N., Dungani, R., Khalil, H. A., Alwani, M. S., et al., Natural weathering studies of oil palm trunk lumber (OPTL) green polymer composites enhanced with oil palm shell (OPS) nanoparticles. *SpringerPlus* 2013, *2*, 592.
67. Morán, J., Vázquez, A., Cyras, V., Bio-nanocomposites based on derivatized potato starch and cellulose, preparation and characterization. *Journal of Materials Science* 2013, *48*, 7196–7203.
68. Gorrasi, G., Pantani, R., Effect of PLA grades and morphologies on hydrolytic degradation at composting temperature: assessment of structural modification and kinetic parameters. *Polymer Degradation and Stability* 2013, *98*, 1006–1014.
69. Hoglund, A., Odelius, K., Albertsson A-C, Crucial differences in the hydrolytic degradation between industrial polylactide and laboratory-scale poly (L-lactide). *ACS Applied Materials & Interfaces* 2012, *4*, 2788–2793.
70. Yatigala, N., Thermal, Physico-Mechanical and Degradation Characteristics of Compatibilized Biodegradable Biopolymers and Composites. MSc diss., North Dakota State University, 2017.
71. Rutkowska, M., Krasowska, K., Heimowska, A., Adamus, G., et al., Environmental degradation of blends of atactic poly [(R, S)-3-hydroxybutyrate] with natural PHBV in Baltic Sea water and compost with activated sludge. *Journal of Polymers and the Environment* 2008, *16*, 183–191.
72. Gilmore, D. F., Fuller, R. C., Lenz, R., Biodegradation of poly (beta-hydroxyalkanoates). *Degradable Materials: Perspectives, Issues and Opportunities*. CRC, Boston 1990, pp. 481–514.

73. Doi, Y., Kumagai, Y., Tanahashi, N., Mukai, K., Structural effects on biodegradation of microbial and synthetic poly (hydroxyalkanoates). *Biodegradable Polymers and Plastics*, edited by M. Vert et al. *Royal Society of Chemistry* 1992, pp. 139–148.

74. Domenek, S., Feuilloley, P., Gratraud, J., Morel, M.-H., Guilbert, S., Biodegradability of wheat gluten based bioplastics. *Chemosphere* 2004, *54*, 551–559.

Chapter 3

Green Composites Reinforced with Cellulose

L. A. Granda, Q. Tarrés, F. X. Espinach, J. A. Méndez, and M. Delgado-Aguilar
LEPAMAP-PRODIS Research Group, University of Girona, C/ Maria Aurèlia Capmany, 61, Girona 17003, Spain
m.delgado@udg.edu

Cellulose reinforced composites are a strong and real alternative to those reinforced with mineral and synthetic fibers in terms of environment, economics, and human safety. This chapter aims to show a broad overview of different types of cellulose reinforcement in terms of morphology, production method, and properties. In addition, the advantages of using micro- or nano-sized cellulose reinforcement are discussed from the production, processability, and application point of view. In this sense, thermoplastic, thermoset, and rubber-based composites will be described in terms of production and applications, as well as their potential use together with micro- and nanostructured cellulosic fibers.

Green Micro- and Nanocomposites
Edited by Sabu Thomas, Abitha V. K., and Hanna J. Maria

ISBN 978-981-4968-79-9 (Hardcover), 978-1-003-42756-8 (eBook)
www.jennystanford.com

3.1 Cellulose: From Micro to Nano

Cellulose, as known today, was discovered almost 200 years ago by the French chemist Anselme Payen, who isolated it from plant matter and determined its chemical formula [1]. Cellulose is an organic compound consisting of a polysaccharide with linear chains of several hundred to many thousand β(1,4) linkages between glucose units and it constitutes the most abundant polymer on Earth. It is present in wood and plants and is mostly used for paperboard and paper production, and also for cellophane, rayon, textile, and fiberboards [2–4]. Besides, conversion from crops to biofuels can be considered one of its potential applications [5]. Cellulose is interesting due to its properties, such as high strength, lightweight, renewability, biodegradability, and thermal stability.

Cellulose can be found in the form of cellulose microfibers, cellulose nanofibers, cellulose nanocrystals (CNC), or even can be grown from bacteria, which is commonly called bacterial cellulose (BC). The first three are always obtained by a top-down approach, while the last one is based on bottom-up methodologies. In this section, each type of cellulose form will be explained in detail, showing the relevance of using one type instead of the other in terms of properties and commercial availability.

3.1.1 Cellulose Microfibers

Fibers can be obtained from different raw materials. Traditionally, fibers have been obtained from wood but there are other sources like plants, forest residues, or agricultural residues that can be used as a raw material. Within wood fibers, we can divide it into softwood and hardwood. The raw material used for fiber achievement plays an important role in fiber properties. There are some properties of the fibers like their different morphology, length and width, chemical composition, and crystallinity limit that condition their use. As is known, fibers are composed of cellulose, hemicellulose, lignin, extractive, and ash. However, a large variety of possible treatments to obtain fibers also drive to different properties of the resulting pulp. This variety of raw material treatments can be classified into mechanical pulp (MP), thermomechanical pulp (TMP), chemical-thermomechanical pulp (CTMP), semichemical pulp (SQ), and chemical pulp (CP). The use of the high amount of chemicals for fiber achievement not only provides fibers with less lignin, extractive contents, and their individualization but also reduces the yield on raw material use.

In general, the users of fibers know that the use of a suitable mix of fibers can achieve the standard requirements for each application on their respective products. In this sense, the fibers unbleached can be used on paper bags or paper packaging production. In that case, producers need fibers with high mechanical properties but the color does not result in an important parameter. Nevertheless, on copy papers it is not enough to have mechanical properties. Also, it is necessary to be smooth, white, and printable. In recent years, one of the rising uses of microfibers has become the substitution of synthetic or mineral fibers [6, 7]. The reinforcement of thermoplastic polymeric matrix with microfibers has been studied in a great number of works [8–10]. Although the use of natural microfibers results in a not trifling enhancement of the mechanical properties of the resulting composite, its strengthening capacity had been shown lower than mineral fibers as glass fibers [11]. However, the use of high percentages of microfibers (30–40%) on composite materials drives an important reduction in the use of nonrenewable resources. Also, their use as a substitute for glass fibers helps in the recyclability of plastics [12]. In short, the use of microfibers in composite production involves a notable environmental benefit.

3.1.2 Cellulose Nanofibers

The study of high fiber individualization has given way to nanocellulose production. Even so, nanocellulose is a general concept. Inside the nanocellulose term, we can find cellulose nanofibers, cellulose nanocrystals, and bacterial cellulose. Cellulose nanofibers (CNF) are characterized by a width range between 3 and 300 nm, more than 2000 nm of length, and less than 70% of crystallinity.

During the 1980s, ITT Rayonier laboratories in New Jersey (USA) studied the named microfibrillated cellulose from the patent developed by Turbak et al., 1985 [13]. The difficulty in nanofibrillation of fibers and the high amount of energy necessary promoted the study of different pretreatments resulting in nanofibrillated cellulose (CNF). Its obtainment is based on three steps. Firstly, fibers are subjected to a pretreatment using one or a combination of different methodologies. Secondly, pretreated fibers are brought to a mechanical desestructuration. Finally, the obtained CNF is characterized.

The use of cellulose nanofibers presents a number of advantages, such as a reactive surface with OH presence, high aspect ratio, transparency, or low density (1.6 g/cm^3). Even so, their high hydrophilicity is necessary to account for their use on composites where matrices usually are hydrophobic.

As in microfibers, cellulose nanofibers can be obtained from softwood, hardwood, annual plants, or agriculture residues. Usually, nanofibers are obtained from bleached microfibers. In the same way as microfibers, the chemical composition of its raw material is also decisive in the nanofibrillation process [14]. As an example, hemicellulose content present on fibers has a very important role [15]. High contents of hemicellulose contribute to helping nanofibrillation process. This phenomenon is related to the amorphous character of hemicellulose and its capacity to bond by hydrogen bonds with fiber surface. It has been proved that as the hemicellulose content of fibers is reduced, the yield of nanofibrillation at the same energy is also reduced [15].

In the same vein, it has been confirmed that a residual content of lignin on fibers reduces the energy necessary for their nanofibrillation [15–18]. In this sense, recent studies demonstrate that by the use of agricultural residues with more than 3% (w/w) of lignin, it is possible to obtain cellulose nanofibers with only mechanical treatments [19].

3.1.2.1 Pretreatments

One of the most important parameters that are possible to control CNF production is the pretreatment methodology. This pretreatment is decisive in order to obtain desired CNF characteristics and its production cost. There are a multitude of models and methodologies that can be applied to pretreat fibers, such as TEMPO oxidation, enzymatic hydrolysis, acid hydrolysis, or mechanical refining.

Oxidation catalyzed by 2,2,6,6-tetramethylpiperidine-1-oxyl (TEMPO) is one of the most important pretreatments, if not the most important on CNF production. This process can be carried on in two ways, at basic pH [20] and neutral pH [21] with different results. TEMPO is a catalyst that acts on the oxidation of 6-carbon (C6) of cellulose chain together with NaClO and NaBr. The process of oxidation of the hydroxyl groups to carboxylic groups, the last ones with higher specific volume, helps the desestructuration [20].

On the other hand, the use of enzymes in the pretreatment of fibers on the cellulose nanofibers is also one of the most used. The pretreatment of enzymatic hydrolysis consists of the addition of endo-beta-1,4-glucanase on the fiber suspension. The function of this endoglucanase is the break of beta-1,4 bonds present on the C1 of glucose and C4 of the following unit of glucose on the amorphous part of cellulose. Usually, this pretreatment is combined with one or two steps of mechanical refining in order to facilitate the desestructuration [14, 22].

Another pretreatment used in cellulose nanofiber production is the use of soft acid hydrolysis. However, this kind of pretreatment is more associated with the production of cellulose nanocrystals due to the reduction of the degree of polymerization [23].

Finally, the mechanical pretreatment consists of the mechanical refining of fibers. This process helps in the desestructuration of fibers by the reduction of fibers length and the separation of fiber bundles. Mechanical pretreatment is usually used in combination with one of the abovementioned pretreatments, but it also can be used alone [24].

3.1.2.2 Desestructuration

For their part, the destructuration stage can be performed by different equipment: high pressure homogenizer, microfluidizer, or grinder.

High pressure homogenizer is the equipment based on the flow of fiber suspension through a compression ring. The fiber suspension is compelled to pass through this ring with reduced dimensions. The forces of friction and impact generate the fiber desestructuration. These friction forces can be associated with the friction between fibers-fibers, fibers-walls of pipe, and fibers-ring perimeter. On the other hand, the microfluidizer is based on the same system but changing the ring of high pressure homogenizer for a pipe zone with a z form. Finally, the grinder is the equipment based on the conventional mechanical refining. In this case, the mechanical refiner has a more sensitive adjust. The grinder produces friction forces between rotor slots, stator, and fibers.

3.1.3 Cellulose Nanocrystals

Cellulose nanocrystals (CNC) are prepared by means of acid hydrolysis intended to dissolve amorphous cellulose chains from fibers. Their geometry is better defined than cellulose micro- and nanofibers due to the crystalline character [25]. As for the preparation of cellulose micro- and nanofibers, several treatments before conducting the main treatment have to be conducted. In this sense, processes like alkaline and organosolv digestion must be conducted on the raw material, as well as a bleaching stage to remove remaining lignin thereof. These treatments must be conducted only in the case when the selected raw material is not pure cellulose, since components such as hemicellulose, lignin, waxes, extractables, and ash can interfere in the hydrolysis process. Afterward, acid hydrolysis can be conducted in two different ways: sulfuric acid hydrolysis (60–70 wt%) and hydrochloric acid hydrolysis (2.5–4 N) in reflux [26]. Different treatment times and acid concentrations can be found in the literature and are well described by Dufresne [27].

Amorphous regions of cellulose are randomly oriented and have a lower bulk density than crystalline domains [28]. These amorphous regions can be considered as susceptible areas to be attacked by the acid and, under controlled conditions, they may be dissolved leading to nano-sized crystals [29]. In recent years, cellulose nanocrystals have received different terminologies, including the following: cellulose micelles [30], level-off D.P. cellulose product [31], whiskers [32], cellulose crystallites [33], cellulose microcrystals [34], cellulose nanocrystals [25, 35], cellulose nanowhiskers [36] and nanocrystalline cellulose [37].

One of the main issues to be solved with the production of CNC is the low yield with respect to the initial amount of the raw material. Several authors have shown the effect of using different raw materials and treatments, but the yield is usually about 15 wt% (about 35–45 wt% with respect to the initial cellulose content). While it is true that Yu et al. obtained CNF with a 93% of yield, there are still missing successful and cost-effective techniques for CNC isolation with high yields [26].

CNC's length is usually of the order of a few micrometers and their width on the order of nanometers. However, nanoparticles resulting from hydrolysis treatment are constituted of a few laterally bound elementary crystallites that are not separated by conventional acid hydrolysis. This morphology can lead to CNC with different aspect ratios, varying from 10 to 70 depending on the treatment severity and raw materials [38].

As in the case of cellulose nanofibers, the mechanical properties and, thus, the reinforcing potential of CNC will vary as a function of many factors, such as the degree of polymerization, specific surface area (bonding capacity), level of sulfation, and crystallinity.

In summary, CNC are produced by dissolving amorphous regions of cellulosic fibers, leading to completely isolated crystals. This dissolving process is commonly conducted by acid hydrolysis (sulfuric/hydrochloric) and leads to nanostructured cellulose-based materials with interesting properties to be used as reinforcement in thermoplastics, thermosets, and even rubbers.

3.1.4 Bacterial Cellulose

Cellulose is commonly extracted from plants and trees. However, there is another way to obtain cellulose with an identical molecular structure to that isolated from flora: synthesis by bacteria. This bacterial cellulose (BC) is obtained from a bottom-up approach, making it completely different in terms of obtaining than that from cellulose fibers. It is a product of microbes' primary metabolic processes, by bacteria from several genera (Table 3.1).

Table 3.1 Bacterial cellulose producers [27]

Genus	Cellulose structure
Acetobacter	Extracellular pellicle composed of ribbons
Achromobacter	Fibrils
Aerobacter	Fibrils
Agrobacterium	Short fibrils
Alcaligenes	Fibrils
Pseudomonas	No distinct fibrils
Rhizobium	Short fibrils
Sarcina	Amorphous cellulose
Zoogloea	Not well defined

Bacterial cellulose presents higher purity than cellulose from plants, which is commonly associated with hemicellulose and lignin. Another advantage is that it can be functionalized from the beginning of the microbial process, leading to well-functionalized fibers. In addition, it can serve as the raw material for CNF and CNC production, leading to purer structures. The structure of BC is finer and more intricate than that of cellulose from plants, and BC fibers present higher tensile strength due to their higher aspect ratio. BC can be oriented during synthesis to obtain uniaxially strengthened membranes. Besides, BC can be genetically modified and can absorb significant amounts of water, the fact that makes them suitable to be used in biomedicine. However, lack of large-scale production, lack of timely expansion, and its high price make BC unsuitable to be used in bulk products, such are composites for structural applications.

Nonetheless, the fact that BC grows from bacteria makes it a good candidate to be used in those matrices in which the interphase with cellulose is something still unsolved. This is the case of polylactic acid, where several efforts are being made to make matrix and fiber surfaces compatible. In this sense, the fact that BC could grow during PLA polymerization opens a window of opportunities for fully biobased and biodegradable composites for high-performance applications.

3.2 Introduction to Cellulose Micro- and Nanocomposites

The differences between the fiber and micro- and nanocellulose ambits are not clearly defined and could vary from one author to other. Nonetheless, it is accepted that cellulose fibers correspond to lengths higher than 2000 μm, microfibrillated cellulose from 0.5 to 10 μm, and nanofibrillated cellulose from 0.5 to 2 μm, with their correspondent diameters from 20 to 50 μm, 10 to 100 μm, and 4 to 20 μm, respectively, but in a lot of cases the micro and nano ambits are united [39–42]. Consequently, when the authors talk about microfibrillated cellulose and nanocellulose, and their composites and composites, most of the time they are referring to the same kind of materials.

Anyhow, the use of any kind of natural fiber has the objective of obtaining a new product with enhanced properties. These enhancements can be focused on mechanical, economic, or environmental properties. In the case of cellulose-based composites, it is possible to attend almost all the cited objectives. Cellulosic and lignocellulosic fibers present comparatively good intrinsic properties, and due to their low density (compared to glass fibers), their relative intrinsic properties could be even more advantageous [43, 44]. The main factors affecting the properties of a cellulosic fiber reinforced composite are the innate variability of the properties of natural fibers, affected by the harvest, type of fiber, origin, treatments; the nature of the matrix, the quality of the interphase, the dispersion of the fibers inside the matrix and its orientation; and the process used to prepare the composite and the porosity of the material [45]. So, such factors center a great quantity of the bibliography at hand. Generally speaking, hemp, ramie, jute, and flax fibers showed the highest intrinsic properties and the highest strengthening and stiffening abilities. Some authors propose some composites based on such fibers as possible substitutes for glass fiber reinforced composites [46].

The research on cellulose nanomaterials started in the 1990s and has increased exponentially up to today with more than 6000 published scientific works [47]. In the case of nanofibers, such intrinsic properties are higher. In that sense, it is possible to obtain almost the same mechanical properties by reinforcing a composite with a 30% w/w of fibers than with a 2.5% w/w of nanofibers. Such high increases in the mechanical properties of the nanocomposites have been reported and some authors referred to the importance of the interphase between the fibers and the matrix [48]. Nonetheless, this behavior is not a clear advantage, as usually matrices are more expensive than natural fibers and, consequently, replacing the higher amount of such matrices with a cheaper phase seems to have more sense from an economic point of view [49, 50]. From an environmental point of view, it seems normal to view a renewable material as natural fibers or cellulosic fibers as more environmentally friendly than oil-based or mineral reinforcement [51–54]. Nonetheless, there is also controversy in that aspect [55–57].

Anyway, cellulosic fibers add additional advantages over mineral fibers, being less abrasive, thus lengthening the lifespan of the machinery, and less harmful when inhaled [58]. There are also some disadvantages, mainly their lower durability and impact strength, high variability and standard deviations of their properties, and great water absorption, which change the mechanical properties.

Regarding the matrices, it is important to signal that cellulose degrades quickly when the processing temperatures are higher than 200°C. This is a limiting aspect when selecting the matrices to reinforce, being the most studied thermoplastic polyolefin as PP or HDPE [45]. The main problem when a hydrophobic polyolefin is reinforced with a hydrophilic reinforcement is obtaining a good interphase. The literature shows a plethora of treatments, but to date, adding low percentages of coupling agents, based on maleated matrices, has proved to be the best method to enhance the mechanical properties of the materials. The amount of fiber contents is limited by the technology used to prepare the specimens or the products. In the case of mold injected materials, it is difficult to find materials with reinforcement contents higher than a 45% w/w.

In the case of thermosets, the most common are epoxy resins, phenol formaldehyde, and polyester.

Moreover, the actual focus on sustainability has increased the use of biobased or biodegradable matrices. In this case, starch-based matrices, polylactic acid, polycaprolactones, and biopolymers are the most present in the literature. Among them, polylactic acid due to its high mechanical properties is one of the most promising options. Its main drawback is its cost in front of commodities like PP or HDPE.

Over the past decades, the interest in cellulose and nanocellulose-based composites has increased. There are nowadays a growing number of scientific articles and books referring to the issue. It is easy to find review articles on the state of the art of such composite materials and also specialized books [45, 47, 59–65], so it is a very lively and growing research field.

3.3 Thermoplastic Cellulose-Based Micro- and Nanocomposites

There are different criteria to classify the polymer composites based on polymers as matrices. Probably the most important, and often useful to understand the final macroscopic properties and processability of the composite, is based on the kind of polymer used as the matrix. In this line, there are three families of composites: thermoplastic, thermoset, and elastomeric composites.

Thermoplastic composites, as the namesake matrices, can be processed by methodologies based on the heating of the material above their melting temperature to allow flow capacity to conform to the final shape after cooling. The most often used thermoplastic polymer matrices to produce cellulosic or lignocellulosic composite materials are polyolefins, where the most representative members are polypropylene (PP) and polyethylene (PE). This does not mean that other thermoplastic matrices have been also considered to be reinforced with cellulosic fibers, although to a lesser extent [66–68]. In Europe, polyolefins production represents almost 50% (48.5%) of the global production of polymers, which suggests that reinforcing or filling such matrices with natural fibers can be perceived as a model of environmental control to decrease their consumption. It must be noted that composites with a high reinforcing percentage reduce the percentage of the polymeric matrix, with the corresponding saving in raw material of a non-renewable resource.

For several decades polyolefins have been reinforced with cellulosic fibers, from those derived from wood (softwood/hardwood), annual plants, and residues (newspaper paper, agro residues, etc.). The aim of this production is guided to produce an alternative material, based on the same polymer matrix, but reinforced with mineral derivative reinforcements: glass fiber, carbon fibers, and aramid fibers. These materials, although presenting excellent mechanical properties, high dimensional stability, and moderate density, compared to metals, also present important drawbacks related to their low recyclability, unhealthy manipulation of the reinforcements (that is the case of glass fibers) [69], and null sustainability. Such kind of limitations can be replayed by means of the substitution of this non-sustainable reinforcement by bioreinforcements derived from cellulose, giving rise to materials with comparable intrinsic mechanical properties, higher recyclable capacity, due to the higher flexibility of these fibers, and lower density. But to acquire such kind of positive character, different modifications must be carried out in order to induce chemical compatibility between both phases. Polyolefins are very hydrophobic materials derived from long hydrocarbon polymer chains, and hydrophilic cellulosic fibers are characterized by the hydroxyl groups supported by C2, C3, and C6 carbons of the cellobiose unit. In this context, there is an important loss of compatibility at the interfacial level, damaging the optimal transfer of loads from the matrix to the reinforcement. Such kind of low compatibility leads to the consideration of these fibers as a filler, as their incorporation into polyolefins does not increase their tensile strength. The way to improve such properties can be considered from two different points of view: physical and chemical. The physical one is based on the purification/cleaning of the surface of the fibers in order to allow the mechanical anchoring of the matrix on the surface of the fiber [70]. This methodology, named mercerization, is based on the treatment of the fibers with an alkali solution, normally NaOH, to remove molecules with low molecular weight and extractables. But even with this cleaning of the surface of the fiber, the different chemical structure of both components does not allow important improvements in tensile strength. So, the chemical adaptation of both surfaces should be performed to increase interaction. To reach such results, very different chemical modifications of the cellulose units on the surface of the fibers have been carried out with very different success outcomes. Procedures based on acetylation [70–72], silanization [73], branching with short polymer chains on the surface [67], and others have been performed. The objective of this modification is the reduction of the polarity of cellulose to be compatible with polyolefins, mainly PP. But the most often used compatibility agent to chemically bond PP and cellulose fibers, and others with high polarity [74–76], is maleated polypropylene (MAH-PP). This agent is based on the chemical anchoring of maleic anhydride in the structure of polypropylene by means of a radical reaction [77]. The incorporation of MAH-PP allows for increasing the ultimate tensile strength by means of the establishment of covalent bonding (ester) between the anhydride functional group and the hydroxyl groups of cellulose. Moreover, the hydrophobic aliphatic fraction of MAH-PP, based on the structure of PP, diffuses through the polymer matrix of the composite by an entanglement mechanism, finishing the coupling.

The addition of nanostructured celluloses to thermoplastics is a very limited way to improve the mechanical properties of the matrices due to the difficulty to incorporate them into the polymer. This reinforcement is obtained as a diluted suspension of nanofibers [20, 78], which must be included in the matrix, being hydrophilic polymers suitable for those requirements. In this sense, starch, poly(vinyl alcohol), and other hydrophilic matrices allow the chemical interactions between their hydroxyl groups and those of cellulose. Some authors have reported important improvements in the mechanical properties of such materials with the target application in the sector of packaging [49].

The incorporation of nanocelluloses (nanofibrillated as well as nanocrystals) to hydrophobic polymer matrices is still under intensive research due to the difficulties to disperse them avoiding the formation of aggregates that limit the transfer of loads from the matrix to the fibers.

3.4 Thermoset Cellulose-Based Micro- and Nanocomposites

The formulation of thermoset cellulose reinforced composites is a result of the mixing of curable resins with cellulose fibers. The low viscosity at a low temperature of the resins helps the incorporation of the fibers into the composite. Thus, it is possible to incorporate fibers in the shape of mats based on unidirectional (UD) and bidirectional (BD) oriented fibers. This methodology allows obtaining an optimal distribution of long fibers into the composite without any other needed equipment to implement fiber dispersion, marking an important difference with respect to the processing of thermoplastic materials, where the dispersion of the fibers into the matrix represents one of the key factors to control determining the success in the composite compounding. Moreover, this methodology also helps to reach values of percentage of reinforcement higher compared to those obtained with thermoplastics. This latter capacity also improves the competitivity of these materials as they are enriched with the most economic component of the composite, the cellulosic fibers.

Thermosetting resins that have been most frequently reinforced with fibers are polyester, vinyl ester, and epoxy resins [79]. From the point of view of reinforcement, such matrices have been reinforced with cellulose fibers mainly derived from annual crops such as flax and hemp [80] (UD/BD). The integrity of the textile structure is mainly maintained by the mechanical interlocking of the fibers [81]. The incorporation of such kind of fiber-like structures into the resins leads to a very important improvement in mechanical properties although they are still lower than those reached by reinforcing with glass or carbon fibers [82]. The lower density of cellulosic fibers offers an almost comparable value of the specific mechanical property [83].

The research in this field has not been limited to the incorporation of cellulosic fibers. An objective of improvement of these results has also been planned to optimize the final mechanical properties. In this sense, different procedures of modification of the fibers, as well as the polymer matrix, have been performed. The most common modification on the surface of the fibers is based again on the cleaning of their surface to remove all the components that impede the fiber-matrix interaction. This cleaning, also named dewaxing, is normally carried out by the treatment of the fibers with boiling ethanol, eliminating waxes and pectins from the surface [84], or with diluted solutions of NaOH [85]. Other conventional methodologies to improve the compatibility of the fiber-matrix system have led to very different results. The chemical modification of flax fibers with methacrylic or propionic anhydride [82] to reinforce unsaturated polyesters led to worse values of tensile stress at break, strain at break, and Young's modulus than the materials reinforced with the unmodified fibers. Such kinds of modifications do not bond chemically the fiber and the matrix but modify the polarity of the surface. On the opposite, there are modifications that improve the mechanical properties of the unmodified reinforcing agent. That is the case of the use of diisocyanates to link chemically the fibers and the polymer matrix. Liu et al. [86] used isophorone diisocyanate (IPDI) to cross-link acrylate epoxidized soybean oil (AESO) with hemp fibers. Moreover, a typical styrene monomer used to copolymerize with curing resin was substituted by *N*-vinyl-2-pyrrolidone (NVP) leading to materials with important improvements in mechanical properties, compared with the material without diisocyanate addition. The incorporation of 5% wt of IPDI into the hemp-AESO composites improved tensile strength (14.9%), tensile modulus (8.4%), flexural strength (19.8%), and flexural modulus (15.9%). Moreover, an increasing percentage of IPDI improved all the mechanical properties at least until percentages of cross-linkers were close to 20%. This increase in strength and stiffness was accompanied by an almost null variation of impact strength, which indicates the high toughness of the network. The incorporation of NVP instead of styrene also decreases the curing temperature and improves the mechanical properties of AESO due to its higher reactivity compared to styrene [87].

Other cross-linkers that have been used to compatibilize cellulose fibers and thermoset resins are zein protein [88], compatibilizing flax fibers and epoxy and vinyl ester resins, and lignin itself [89], decreasing the hydrophilicity of the fibers to improve compatibility with unsaturated polyesters and vinyl esters.

But the development of improved thermoset composites reinforced with cellulose has been based on not only the incorporation of new cross-linkers, but also the synthesis of new resins friendlier with the environment based on renewable resources. Thus, the use during the last years of AESO resins, as mentioned above, poly(lactic acid) cross-linked with pentaerythritol [90], oleic sunflower oil-derived resins [91], ferulic acid-derived esters and amides [92], and other biobased epoxy and polyester resins [93], has allowed a change of philosophy in the fabrication of thermoset materials and their corresponding composites, and together with the cellulosic reinforcement, it is possible to build 100% renewable materials.

From the point of view of cellulose nanoreinforced thermoset composites, there is not so much related literature due to the high difficulty of dispersion of the hydrophilic nanostructured reinforcements inside hydrophobic resins, being cellulose nanocrystals (CNC) the often used. To solve such a situation, a different synthesis of epoxy thermoset resins with higher hydrophilicity has been carried out. Thermoset resins based on the epoxy structure modified with ethylene glycol (side chains) and aliphatic amines were effectively reinforced with different percentages of nanowhiskers of cellulose in the range of 0.5–8% wt [94]. The dispersion of the nanowhiskers was derived from the interaction of the hydroxyl groups of the resin and the fibers by means of the establishment of hydrogen bonding. On the other hand, the nanoreinforcement can be the phase suitable to be modified to compatibilize with hydrophobic resins. That is the case of the surface modification of CNC with a commercial surfactant: phosphoric ester of polyoxyethylene nonylphenyl ether to compatibilize with oil-based resins [91, 95].

3.5 Rubber Cellulose-Based Micro- and Nanocomposites

There are different causes to incorporate filler/reinforcements to elastomeric materials derived from natural rubber or other elastomers based on a synthetic formula.

One of the most important fillers added to natural rubber in order to develop engineering materials is black carbon. However, this filler also confers some electrical conductivity to the materials, being a handicap for applications with electrical isolation demands. To improve such isolating character, other carbon derivatives, such as medium thermal carbon blank, have been effectively incorporated into natural rubber to decrease electrical conductivity due to its higher resistivity. Together with this low electrical conductivity, this reinforcement also confers black color to the pieces and is derived from a mineral nonrenewable origin. In this sense, there is a lot of research focused on alternatives to black carbon as a reinforcing agent of rubber [96]. Cellulose is one of the most promising substitutes due to its natural origin, low capacity for electrical conductivity, thermal capacity, and good mechanical behavior. In fact, it is possible to improve the mechanical strength of elastomeric composites by incorporating up to 15 phr of cellulose, although the strain at break is decreased as expected. Such reinforcing level is similar to that obtained by reinforcing NR with carbon black or thermal carbon black in the range of 20–50 phr. Moreover, the incorporation of cellulose fibers also decreases the vulcanization time, increasing cross-linking due to the polymer-filler interactions [97].

Table 3.2 Elastomeric materials and usual reinforcements based on cellulose

Polymer matrix	Reinforcement	Range of reinforcement	Observations	Ref.
Natural rubber	Cellulose II xanthate	Up to 30 phr	Increase in torque, hardness shore	[98]
XSBR	CNC	Up to 15 phr	Increase in torque, storage and loss moduli, hardness shore	[99]
Natural rubber	CNF (steam explosion)	Up to 10 phr	Improvement in mechanical properties. Reinforcement accelerates biodegradation	[100]
Natural rubber	CNC	Up to 6 phr	Improvement in mechanical properties. Decrease in deformation capacity	[101]
Natural rubber	CNF	Up to 15 phr	NFC influences electrical conductivity	[102]
SBR foams	CNC	Up to 15 phr	Improvement in mechanical strength without damage on strain at break	[103]
SBR	CNC	Up to 20 phr	Improvement in mechanical strength, torque, and Young's modulus	[104]
Oxidized natural rubber	CNC	Up to 5 phr	Oxidation decreases the onset degradation temperature. CNC prevents film disruption in contact with organic solvents	[105]

Natural rubber can be effectively dispersed in a water solution to obtain the corresponding emulsion (latex). This is the usual way of finding natural rubber in nature (*Hevea brasiliensis*). As in the case of the hydrophilic polymer matrices (starch, PVA, and others), it is possible to include the nanocellulose's hydrogel in the latex, and after solvent casting the corresponding NFC reinforced natural rubber can be obtained. Owing to this characteristic, elastomers with the capacity for the formation of water emulsions are suitable materials to be effectively reinforced with nanocellulose.

Table 3.2 summarizes the most often reinforced rubbers as well as the corresponding reinforcement and the most remarkable results.

3.6 Processing of Green Cellulose Composites

All short fiber composites can be processed with the same technology as monolithic polymers are processed. However, the higher viscosity of composites, caused by the addition of a discontinuous phase, leads to applying some modifications in the process.

In thermoplastics, the processing techniques related to composite materials could be separated into two subjects: composite production and composite transformation process. On thermosets and rubbers, there is a previous mixing stage and thereafter the molding and cross-linking process.

3.6.1 Composite Production

Frequently, either for obtaining a good dispersion of the reinforcement within the matrix or for economic criteria, the transformation process is partially separated from the transformation or molding process in which the final product is obtained.

3.6.1.1 Thermoplastic matrix composites

Thermoplastic compounding on an industrial scale usually involves a process of extrusion. In general terms, it can be classified into single-screw extrusion and twin-screw extrusion. The inclusion of distributive elements such as "Saxton," "pineapple," or "Twente" mixers and dispersive elements such as "Barrier," "Maillefer," and "Maddock" mixing elements allows for enhancing the distribution of reinforcements in the matrix by single-screw extrusion. Some examples in the literature can be found where this equipment has been used for compounding, especially for inorganic fillers

[106–108]. However, the single-screw extruder does not disperse fibrous reinforcements as effectively as twin extruders do.

Therefore, twin extrusion is the main industrial process to compound. There can be found two different types of twin extruders whether the turning direction is the same or not: co-rotating extruders and contra-rotating extruders. While contra-rotating extruders apply higher shear strength to the compound, the co-rotating extruders have a higher residence time in the extruder, which leads to a higher dispersion of the discontinuous phase within the matrix. Besides, the lower shear stress might contribute to better morphology ratio conservation, as the fibers suffer lower shrinkage processes.

3.6.1.2 Thermoset and rubber matrix compounding

In thermoset polymers and rubbers, the resin is cured during the processing. Therefore, the mixing of the thermoset or rubber with the fibers or nanofibers takes place in a liquid-solid mixing step. While for mixing microfibers the use of common mixers is frequent, their use for nanofibers dispersion is not adequate and sonication mixers are extensively used.

3.6.2 Composite Transformation

3.6.2.1 Thermoplastic matrix composites

3.6.2.1.1 Extrusion

Extrusion is the act of pressing out through an opening in order to give shape. Therefore, the extrusion processes allow obtaining products with a constant section such as different kinds of profiles, sheets, films, wires, and pipes. This makes extrusion one of the most used processing techniques [109, 110].

3.6.2.1.2 Extruder machine parts

All extruders are constituted by a gearbox, a control panel, a feeder, a barrel and a screw, heaters and thermocouples, and a die. The die is a complement of the machine itself and is fitted together with the extruder. The die is responsible for shaping the material into the final required product.

1. Gearbox: Its function is to apply a force to the screw for turning it. Its power will depend on the screw dimensions and the extruder capacity.
2. Control panel: It allows the modification of the entrance parameters (extrusion temperatures and screw speed) and also displays the answer parameters of the process (melt temperature, torque, and melt pressure before the die).
3. Hooper: Its main role is to keep material feeding constant throughout the process. For composites with high content of reinforcement, the use of hoppers subjected to constant vibration is recommended.
4. Barrel: It provides one of the needed surfaces for submitting the materials to shear strength and also is the element that transfers the heat supplied by the heaters and also removes the heat excess when the temperature is too high. The ratio between the length and the diameter (L/D) is their main design parameter. High L/D screws lead to better plastification and thus high productivity. However, it also increases the residence time of the material inside the barrel. Hence, polymers susceptible to degradation should not be processed with those kinds of barrels.
5. Screw: It is the most important part of the equipment as it is responsible for the transportation of the material inside the barrel, plasticizing it, and supplying pressure to allow the flow of the material through the die. It could be divided into three different zones: the feeding, the compression, and the metering zones. The functions of the feeding zone are to transport the material through the barrel, heat it by the heaters, and allow the exit of the air through the feeding. The compression zone is necessary for adapting the screw to the volume that the material occupies when passing from the solid to the melted phase. In this section, there will be found both solid and melted materials, and the material heating happens by shear stress and by the external heaters. The volume compressed between the two first pitches divided between the volumes compressed between the two final pitches from the screw is called the compression ratio. This is the most important design parameter of a screw. Finally, in the metering zone, the

heating is mainly due to shear stress forces and it supplies the needing pressure to push the material through the breaker plate and the die. The configuration of those parts can differ depending on the material's characteristics. In general terms, amorphous polymers will need a screw with a short feeding zone, a long compression zone, and a moderate compression ratio. Meanwhile, semicrystalline polymers should be processed with screws with long feeding zones and short compression zones with a high compression ratio.

6. Breaker plate: The breaker plate is a metal disk with several holes. It is perpendicularly located to the material flow at the exit of the barrel. Its main purpose is to straighten out the flow, as the turning inertia of the material inside the extruder may lead to defects in the final product [111].
7. Heaters: They supply thermal energy to ease the melting of the polymer inside the extruder.
8. Thermocouples: They are responsible for the temperature measurement in every part of the barrel.

3.6.2.1.3 Differences when extruding cellulose composites

The most important consideration that should be taken into account is temperature processing. The cellulose fiber degradation is comprehended around 200–230°C. This fact limits the processing temperature to that temperature. In the same way, the matrix selection is important, as its melting temperature should be lower than the reinforcement degradation temperature.

Another important fact is related to the screw selection. For the high content of fibrous reinforcement, a screw with a lower compression ratio than a non-reinforced matrix is preferable, as the cellulose fiber is not sensible to compression.

3.6.2.1.4 Injection

Injection molding of thermoplastic materials consists of introducing a melted thermoplastic plastic into a mold that will give shape to the final product. It allows the production of relatively thin products with complex geometries.

The process is a semicontinuous production of thermoplastic pieces subjected to a cycle that is repeated until the production is reached. The cycle can be divided into injection, feeding, cooling, and ejection. Thereafter, the cycle restarts and goes forth.

3.6.2.1.5 Injection machine parts

A standard injection machine is formed by a plastification unit, the mold clamping unit, the gearbox, and the control unit. The plastification unit, very similar to the extruder machine, is formed by a barrel, a screw, and different heating zones and thermocouples.

1. Gearbox: It will transform electrical energy into the mechanical energy needed to move the plastification unit, the screw, and the mobile parts of the mold.
2. Control unit: The process allows different control parameters such as profile temperature of the different zones, feeding volume, injection pressure, injection speed, holding pressure, plastification pressure, and the speeds at which the mold opens and closes.
3. Hopper: Owing to the great importance of a constant supply of the material to the machine, the hopper is a significant element of the machine. As in extrusion, hoppers with vibrational systems are recommended for cellulose composite processing.
4. Barrel: In its interior, there is the screw and supplies one of the surfaces needed for plasticizing the materials. Again, it should have a good heat transmission coefficient for either transferring the heat generated by the heaters to the material or removing its excess. The L/D ratio will be related to the plasticizing efficiency and thus, again, to the productivity of the machine.
5. Screw: Whereas the use of mixing elements is so frequent in extrusion, it shows lower importance for injection molding. Industrially, it is much more frequent the use of masterbatches or pellets containing mineral reinforcement than the proper incorporation of the discontinuous phase during the process. Therefore, while the injection screws are subjected to the same design parameters (compression ratio, etc.) as extrusion, they are frequently simpler than extruders. The screw in injection molding has two different movements: a rotational movement and a longitudinal movement for pressing the material. In

order to prevent the material from going back during the injection, an anti-return valve is set at the end of the screw.

6. Nozzle: It is the part that stays in contact with the mold's sprue bushing. A good contact between the mold and the nozzle is crucial for assuring no material loss through that point. This would lead to demolding problems that would drastically reduce the process productivity.
7. Heaters: They provide heat to the material in order to facilitate the plastification of the materials.
8. Thermocouples: They measure the temperature of each zone of the barrel.

   The mold will be formed by two halves: one of which is fixed and the other one is mobile in order to eject and recover the injected product. For that reason, the mold clamping unit is formed by, at least, a fixed plate, a moving plate, and the mold.
9. Fix plate: It is the part of the injection machine that holds the static half of the mold and the one that first receives the plasticized material from the injection machine.
10. Moving plate: It is the part of the injection machine where the moving part of the mold is assembled.
11. Mold: The mold is the element that confers the shape of the final product. It is subjected to the injection machines through the plates. The plane formed by the two halves of the mold is called the partition plane. The place where it is set is an important issue. The mold is formed by sprues and gates that drive the plastic flow to the cavity where the material takes the shape of the final product. It has also an ejector system to easily remove the injected piece from the mold. There are hot and cold runner molds. For the injection of cellulose reinforced composites, it is inadvisable to use hot runner molds, as the risk of damaging the runner is high. The molds also have a refrigeration system for avoiding mold overheating, which would lead to a productivity reduction [110].

3.6.2.1.6 Differences when injecting cellulose composites

Once again, the low thermal stability of cellulose and wood components such as lignin and hemicellulose is the main drawback of processing those composites by injection molding. Adding those kinds of reinforcements will increase the materials' viscosity. Altogether, the processing conditions should balance the processing conditions of a material with a high viscosity, if compared with monolithic polymers, and a low transformation temperature for not degrading the reinforcement.

Beforehand, the use of matrices with a high melt flow index is recommended. Also, the use of nozzles with a wide hole is urged. In order to compensate for the low MFI of our materials at the processing temperature, it will be needed to increase the injection pressure for fulfilling the mold cavity.

By extrusion and injection molding, it is possible to process polymers reinforced with micro-reinforcements as well as nanoreinforcements [66, 112–114]. On the other hand, by thermoforming and rotational molding, it will only be possible to process polymers reinforced with nanoreinforcements. This happens because of the low strain and the high viscosity of those composites.

3.6.2.1.7 Thermoforming

In thermoforming, the raw material is a previously extruded sheet. In general terms, thermoforming consists of heating that sheet to a rubbery state and then applying elastic deformations before cooling and demolding. The product's wall thickness is not regular, as the parts subjected to higher deformations will have a high thickness reduction.

The process can be divided into three parts: loading (and unloading), heating, and molding. The loading will be dependent on the thickness of sheet. In that sense, thin gage materials are typically fed by roll feed through a continuous process. For thicker materials, the loading process is a semicontinuous one, in which materials are set manually or automatically in the machine. Those sheets are pressed in a clamping frame in order to avoid twists, warping, or movements during the process. The heat transference can be due to radiation, convection, and conduction. The heating of the sheet surface is usually performed by radiation through calrod heaters, ceramic or quartz heaters, combustion heaters, and so on. The sheet core is heated by conduction. Hence, the sheet's thickness and the typical conductivity of the material are important factors to be taken into consideration for assuring the needed time before molding. The molding process can be used in different ways to apply deformation to the sheet: mechanical forces, atmospheric pressure (by applying vacuum on the cavity between the mold and the sheet), pressure force, or a combination of them. When the atmospheric pressure is not enough for forming the sheet to the desired shape, pressure forming may be used. In that case, the practical limit of pressure would be about 10 bars, 10 times bigger than the vacuum forming one. The combination of mechanical forces with vacuum, for instance, may be used for reducing the wall thickness variation [115].

3.6.2.1.8 Rotational molding

Rotational molding is a process that allows the production of hollow pieces without pressure using low investment equipment [110]. Polyolefin is the most frequently used polymer family for that processing. Specifically, more than 80% of the total rotational molding production is from polyethylene [116]. The plastic in that process must be supplied in powder. It consists of the following steps: load, heating, cooling, and discharge. The load is characterized by introducing the needed quantity for producing the required product into a mold. Thereafter, the mold is heated and, at the same time, it is twirled in two axes. Thus, the material is melted and covers the whole mold cavity. Afterward, the mold is cooled, but the twirling is still maintained in order to allow the plastic to turn solid before demoulding. In this process, water sprinklers can be used. Finally, the product is demolded and finished, removing the burrs for instance.

Traditionally, the main applications of that process are toys, tanks for industrial applications, containers, automotive articles, and so on. Compared with blow molding, it is a very inexpensive process with a better wall uniformity that produces products with low residual stresses. However, it is very difficult to work with fibers and no partially oriented molecules are obtained [116].

3.6.2.2 Thermoset matrix composites

Despite the allowance of measuring the curing, or vulcanization, time by rheometers, calorimetric analysis, and so on, the determination of the required molding time is complex [117]. The use of complex models that, besides the kinetic reaction itself, takes into consideration the material's conductivity and the shape of the product [118] is needed for a good mold design.

Although there are other transformation processes for thermoset composites such as pultrusion, or resin transfer molding [119], they have been disregarded in this chapter as they are typical for long fiber composites. Therefore, compression molding, transfer molding, injection molding, and reactive injection molding will be described below.

3.6.2.2.1 Compression molding

It consists of firstly introducing the needed volumetric quantity of mixture that allows for filling of the mold cavity. The mold is placed between the two plates of a press, which can be hydraulically or mechanically activated [117].

Generally, the mold is already heated when the material is fed into the mold. This offers a productivity advantage for serial production regarding feeding the mold when cold. Placing the raw material for assimilating the final product form can reduce the flowing movements of the material and the apparition of air occlusions. Degasification processes are performed in order to remove trapped air. They consist of applying the curing pressure and interrupting it for short moments. Immediately, the pressure is increased again and the reticulation time is measured from that time to the mold opening. Once the curing time has passed, the press is opened. If the molds are attached to the press, this will open the mold. Otherwise, the mold will have to be removed and opened apart. Afterward, the product is demolded and the mold refilled with raw material for the following production. Finally, the burrs generated by the overflowing of the material are removed [117].

3.6.2.2.2 Transfer molding

In transfer molding, the resin is injected into the mold by an injection ram. The simplest installation for transfer molding is a three-piece mold in which the cavity is placed between inferior and intermediate pieces. The latest also contains the transference cavity where the material is fed and heated until softened [120]. The third part of the mold, the transference ram, is responsible for applying pressure into the transference chamber and thus filling the mold's cavity [117]. Like in injection molding, a holding pressure is applied for accommodating sample shrinkage during the curing.

Despite there being cases of filled resins transformed through transfer molding, some considerations must be taken into account. Firstly, the resins must be heated to a temperature that drives to low viscosity, but it should avoid the curing effect [120]. Besides, the diameter in the injection point is uncommonly higher than 0.5 mm. This can lead to obstructions when using cellulose fibers. Therefore, using molds with wider injection points is recommended.

3.6.2.2.3 Injection molding

There are two different injection processes related to thermosets: thermoset injection molding and reactive injection molding. The second one is mainly used for polyurethane systems and the first one is widely used for both polyester and elastomer transformation [120, 121].

In spite of some differences, thermoset injection molding shows some similarities with thermoplastic injection molding. In this case, the injection unit plasticizes the polymer and the mold is always at a higher temperature than the injection unit. Thus, the polymer is cured or vulcanized inside the mold.

Essentially, there are two different kinds of injection units: those in which the plastification and injection are performed by the screw inside the barrel, exactly as in thermoplastic injection; and those in which the injection is performed by an injection ram. In the latter, the plastification takes place in an extruder that feeds an injection chamber. The material stocked in that chamber will be injected into the mold by the injection ram through a nozzle [121].

The injection process is similar to thermoplastic injection, with an injection pressure and a holding pressure. The time in which the machine stays in contact with the mold is a drastic parameter. As it must be enough for letting the material into the entry of the mold to be cured, but low enough to avoid the curing of the material placed in the nozzle.

Reactive injection molding is mainly used for thermosetting polymers which have a high lifetime at low temperatures, but a low curing time when heated. Although polycaproamide, unsaturated polyesters, and epoxy resins can be processed through reactive injection molding (RIM), the processing of polyurethane (PU) systems is the most successful [122]. Despite the process being practically limited to PU, the presence of cellulose nanofibers and cellulose nanocrystals reinforced PU matrices in the bibliography [104, 123, 124] makes that processing method a noteworthy one for being mentioned in that chapter.

The components are mixed in turbulent mixing chambers to assure a good contact between the two reagents [120]. Afterward, they are injected into the mold to be conformed to the product's final shape. The mold is frequently set as an independent assembly. As all the polymerization reactions are exothermal, the mold is not needed to be heated except at the beginning in some cases, for example, in the reactive molding of polycaproamide. The low viscosity of those resins makes unnecessary the use of high pressures. Thus, the mechanical requirements of those molds are not as high as in thermoplastic injection. Therefore, the use of aluminum or even reinforced plastic molds is frequent [122].

The reactive injection of composites is a developed technology either for short fibers or woven or unwoven long fibers. At the latest, the process is similar to the resin transfer molding transformation. On the other hand, the short fibers should be incorporated into one of the reagent mixtures [122]. In the case of cellulose fibers and nanofibers, as their hydroxyl groups can react with isocyanates, the micro- or nanofibers should be added to the polyol reagent's mixture.

3.7 Applications of Cellulose-Based Micro- and Nanocomposites

The natural fiber reinforced composites passed from an opportunity to a reality in the past decades. In that sense, there are some examples in the automotive, building, and product design fields [125].

The automotive and transport industry has been searching for two main but sometimes opposed goals, on the one hand, the increase of the security of the vehicles. This goal has been achieved with the inclusion of every day more and more sophisticated active and passive systems that had redounded in a diminution of the severity of the lesions in the accidents. Nonetheless, the inclusion of such mechanisms has increased the weight of the vehicles, ballasting the second goal. This second goal is the diminution of combustible consumption. It is clear that heavier vehicles will consume more; thus, the use of lighter materials is mandatory to obtain the same security with a lower environmental impact due to CO_2 emissions. As an example, the use of natural reinforced door panels allowed the Mercedes SL a reduction of 20% in weight [125]. In that sense, the use of natural fiber reinforced composites proved particularly interesting. There are some studies on the replacement of glass fiber reinforced matrices with wood fiber reinforced composites [126]. The study proposes the fabrication of some internal panels with lignocellulosic reinforced PP. The authors ensure that using greener composites does not compromise the initial specifications. Natural fiber reinforced composites are commonly used for the front door, rear door and boot liners, parcel shelves, sunroof interior shields, and headrests [125]. As an example of use, the BMW 3, 5, and 7 series use door panels, headliner panels, boot lining, seat backs, noise insulation panels, and door trims manufactured with natural fiber reinforced composites.

Another field interested in the adoption of lighter and environmentally friendlier materials is the building industry. Some countries have started to formulate legal policies related to the life cycle analysis of the buildings and their impact on the environment [127]. Consequently, the interest in biobased materials is increasing. There are studies on the superior sound insulation power of natural fiber reinforced composites [128]. There is interest in the use of the composites for plan façades and insulating materials [129], aiming to greener buildings with facades that could have a positive influence on the hydrothermal conditions of the edifice.

The main use of natural fiber-based composites for building materials is centered on the use of composite panels, roofing sheets, door shutters, frames, and shuttering plates [130].

Another field where more sustainable materials are being seen as an opportunity to create value is the industrial design. Industrial design allows adding value to a product by adding use, function, or perception assets. The designers show their interest in greener products as a social demand, allowing the incorporation of a more socially responsible design [131, 132].

Some products such as trays, water pump bodies, or chairs could be redesigned to allow their manufacture with greener composites [46, 133, 134]. In any case, the use of such composites does not compromise the aspect, functionalities, or perceived quality of the final products, but adds the environmentally friendly label as a feature.

There are a high number of commercially available products manufactured with natural fiber reinforced composites [135]: The Be.e electric scooter by van.eko has a body manufactured with a flax reinforced bio-resin. Focal designed a new speaker's high-end line with a flax fiber cone.

The use of natural fiber reinforced composites is open to a great quantity of products, and it is only weighted by its nowadays comparatively higher price. Its use could dramatically increase in the future due to legal regulations, a decrease in the price of the biomatrices, or a change of perception in the society, adding more value to sustainability.

3.8 Conclusion and Perspectives

Cellulose-based composites have shown their versatility. In the shape of fibers or short fibers, it is possible to obtain composite materials with mechanical properties comparable to glass fiber reinforced materials. These composites have overcome the difficulties of achieving a good reinforcement–matrix interface, and nowadays their mechanical properties compare well with those of commercial materials. On the other hand, the cellulose-based composites have shown a good adaptation to the traditional manufacturing and production technologies, being in some cases better adapted due to their lesser abrasiveness. Cellulose fibers are also less harmful to the workers. While the reinforcement of polyolefin and some thermosets is almost a mature technology, the adoption of biobased matrices is showing great advances toward fully biobased composite materials. In that sense, one of the better positioned matrices is the polylactic acid. On the negative side, there are some aspects such as the water absorption of the cellulosic fibers and its effect on the mechanical properties of the fibers that need more research to be successfully solved. There is also more research needed to know its dynamic and longtime behavior.

The nanocellulose or microfibrillated cellulose-based composite materials show even more promising opportunities. Owing to its unique nature, it is possible to obtain functional materials with properties that could not be achieved through the incorporation of common cellulosic fibers. The major concerns of the nanocellulose fibers are the difficulty of achieving good dispersion. The problem has been partially solved by using thermoset impregnated nanopapers and thus high content, high properties materials. In some cases, it has been proved that comparatively short percentages of nanofibers produced composite materials with better properties than those in the market. Nonetheless, nowadays, the biggest nanocellulose drawback is its cost. Such cost restricts the nanocellulose-based composites to compete with glass fiber-based composites and limits their use to high-value added products.

Cellulose-based green composites are a great opportunity to create new materials with enhanced socially responsible properties. The literature shows that these composites have, nowadays, good enough properties to compete with glass fiber-based composites. Nonetheless, there is a social barrier to overcome. The product value, based only on a monetary basis, affects negatively the use of more expensive materials, whatever their environmental value. Science has shown the opportunities, and now the market must be able to identify such opportunities and convert them to value.

References

1. Payen, A. (1838) Mémoire sur la composition du tissu propre des plantes et du ligneux. *Comptes Rendus*, **7**, pp. 1052–1056.
2. Mazhari Mousavi, S.M., et al. (2013) Papermaking potential of rapeseed straw, a new agricultural-based fiber source. *Journal of Cleaner Production*, **52**, pp. 420–424.
3. Mancera, C., et al. (2012) Feasibility of incorporating treated lignins in fiberboards made from agricultural waste. *Waste Management*, **32**(10), pp. 1962–1967.
4. Mwasha, A. (2009) Designing bio-based geotextiles for reinforcing an embankment erected on the soft soil. *Materials & Design*, **30**(7), pp. 2657–2664.

5. Yang, B. and C.E. Wyman (2008) Pretreatment: the key to unlocking low-cost cellulosic ethanol. *Biofuels, Bioproducts and Biorefining*, **2**(1), pp. 26–40.
6. Granda, L.A., et al. (2016) Semichemical fibres of Leucaena collinsii reinforced polypropylene composites: Young's modulus analysis and fibre diameter effect on the stiffness. *Composites Part B*, **92**, pp. 332–337.
7. Naghmouchi, I., et al. (2015) Polypropylene composites based on lignocellulosic fillers: how the filler morphology affects the composite properties. *Materials & Design*, **65**, pp. 454–461.
8. Codispoti, R., et al. (2015) Mechanical performance of natural fiber-reinforced composites for the strengthening of masonry. *Composites Part B*, **77**, pp. 74–83.
9. Jiménez, A.M., et al. (2016) Tensile strength assessment of injection-molded high yield sugarcane bagasse-reinforced polypropylene. *BioResources*, **11**, pp. 6346–6361.
10. Karmaker, A.C. and J.A. Youngquist (1996) Injection molding of polypropylene reinforced with short jute fibers. *Journal of Applied Polymer Science*, **62**, pp. 1147–1151.
11. López, J.P., et al. (2012) Tensile strength characteristics of polypropylene composites reinforced with stone groundwood fibers from softwood. *BioResource*, **7**, pp. 3188–3200.
12. Lopez, J.P., et al. (2012) Recycling ability of biodegradable matrices and their cellulose-reinforced composites in a plastic recycling stream. *Journal of Polymers and the Environment*, **20**, pp. 96–103.
13. Turbak, A.F., et al. (1983) Suspensions Containing Microfibrillated Cellulose, US Patent No. 4378381, 1983.
14. Pääkko, M., et al. (2007) Enzymatic hydrolysis combined with mechanical shearing and high-pressure homogenization for nanoscale cellulose fibrils and strong gels. *Biomacromolecules*, **8**, pp. 1934–1941.
15. Chaker, A., et al. (2013) Key role of the hemicellulose content and the cell morphology on the nanofibrillation effectiveness of cellulose pulps. *Cellulose*, **20**, pp. 2863–2875.
16. Ferrer, A., et al. (2012) Valorization of residual empty palm fruit bunch fibers (EPFBF) by microfluidization: production of nanofibrillated cellulose and EPFBF nanopaper. *Bioresource Technology*, **125**, pp. 249–255.

17. Ferrer, A., et al. (2012) Effect of residual lignin and heteropolysaccharides in nanofibrillar cellulose and nanopaper from wood fibers. *Cellulose*, **19**, pp. 2179–2193.
18. Rojo, E., et al. (2015) Comprehensive elucidation of the effect of residual lignin on the physical, barrier, mechanical and surface properties of nanocellulose films. *Green Chemistry*, **17**(3) pp. 1853–1866.
19. Espinosa, E., et al. (2015) Suitability of wheat straw semichemical pulp for the fabrication of lignocellulosic nanofibres and their application to papermaking slurries. *Cellulose*, **23**(1), pp. 837–852.
20. Saito, T., et al. (2007) Cellulose nanofibers prepared by TEMPO-mediated oxidation of native cellulose. *Biomacromolecules*, **8**, pp. 2485–2491.
21. Besbes, I., S. Alila, and S. Boufi (2011) Nanofibrillated cellulose from TEMPO-oxidized eucalyptus fibres: effect of the carboxyl content. *Carbohydrate Polymers*, **84**, pp. 975–983.
22. Henriksson, M., et al. (2007) An environmentally friendly method for enzyme-assisted preparation of microfibrillated cellulose (MFC) nanofibers. *European Polymer Journal*, **43**, pp. 3434–3441.
23. Zimmermann, T., E. Pöhler, and T. Geiger (2004) Cellulose fibrils for polymer reinforcement. *Advanced Engineering Materials*, **6**, pp. 754–761.
24. Gamelas, J.A.F., et al. (2015) On the morphology of cellulose nanofibrils obtained by TEMPO-mediated oxidation and mechanical treatment. *Micron*, **72**, pp. 28–33.
25. Habibi, Y., L.A. Lucia, and O.J. Rojas (2010) Cellulose nanocrystals: chemistry, self-assembly, and applications. *Chemical Reviews*, **110**(6), pp. 3479–3500.
26. Yu, H., et al. (2013) Facile extraction of thermally stable cellulose nanocrystals with a high yield of 93% through hydrochloric acid hydrolysis under hydrothermal conditions. *Journal of Materials Chemistry A*, **1**(12), pp. 3938–3944.
27. Dufresne, A., *Nanocellulose: From Nature to High Performance Tailored Materials*. 2013: Walter de Gruyter.
28. Saxena, I.M. and R.M. Brown (2005) Cellulose biosynthesis: current views and evolving concepts. *Annals of Botany*, **96**(1), pp. 9–21.
29. Thielemans, W., C.R. Warbey, and D.A. Walsh (2009) Permselective nanostructured membranes based on cellulose nanowhiskers. *Green Chemistry*, **11**(4), pp. 531–537.
30. Rånby, B.G. (1951). Fibrous macromolecular systems. Cellulose and muscle. The colloidal properties of cellulose micelles. *Discussions of the Faraday Society*, **11**, pp. 158–164.

31. Battista, O.A. and P.A. Smith, *Level-off DP cellulose products*. 1961, U.S. Patent.
32. Dufresne, A. (2008) Polysaccharide nano crystal reinforced nanocomposites. *Canadian Journal of Chemistry*, **86**(6), pp. 484–494.
33. Dong, X.M., et al. (1996) Effects of ionic strength on the isotropic-chiral nematic phase transition of suspensions of cellulose crystallites. *Langmuir*, **12**(8), pp. 2076–2082.
34. Araki, J. and S. Kuga (2001) Effect of trace electrolyte on liquid crystal type of cellulose microcrystals. *Langmuir*, **17**(15), pp. 4493–4496.
35. Bondeson, D., A. Mathew, and K. Oksman (2006) Optimization of the isolation of nanocrystals from microcrystalline cellulose by acid hydrolysis. *Cellulose*, **13**(2), pp. 171–180.
36. Kvien, I. and K. Oksman (2007) Orientation of cellulose nanowhiskers in polyvinyl alcohol. *Applied Physics A*, **87**(4), pp. 641–643.
37. Peng, B.L., et al. (2011) Chemistry and applications of nanocrystalline cellulose and its derivatives: a nanotechnology perspective. *The Canadian Journal of Chemical Engineering*, **89**(5), pp. 1191–1206.
38. Boluk, Y., et al. (2011) Suspension viscosities and shape parameter of cellulose nanocrystals (CNC). *Colloids and Surfaces A: Physicochemical and Engineering Aspects*, **377**(1), pp. 297–303.
39. Moon, R.J., et al. (2011) Cellulose nanomaterials review: structure, properties and nanocomposites. *Chemical Society Reviews*, **40**(7), pp. 3941–3994.
40. Klemm, D., et al. (2005) Cellulose: fascinating biopolymer and sustainable raw material. *Angewandte Chemie International Edition*, **44**(22), pp. 3358–3393.
41. Dufresne, A., J.-Y. Cavaillé, and M.R. Vignon (1997) Mechanical behavior of sheets prepared from sugar beet cellulose microfibrils. *Journal of Applied Polymer Science*, **64**(6), pp. 1185–1194.
42. Iwamoto, S., K. Abe, and H. Yano (2008) The effect of hemicelluloses on wood pulp nanofibrillation and nanofiber network characteristics. *Biomacromolecules*, **9**(3), pp. 1022–1026.
43. Reixach, R., et al. (2013) Micromechanics of mechanical, thermomechanical, and chemi-thermomechanical pulp from orange tree pruning as polypropylene reinforcement: a comparative study. *Bioresources*, **8**(3).
44. Vallejos, M.E., et al. (2012) Micromechanics of hemp strands in polypropylene composites. *Composites Science and Technology*, **72**(10), pp. 1209–1213.

45. Pickering, K.L., M.A. Efendy, and T.M. Le (2016) A review of recent developments in natural fibre composites and their mechanical performance. *Composites Part A: Applied Science and Manufacturing*, **83**, pp. 98–112.

46. Serrano, A., et al. (2014) Study on the technical feasibility of replacing glass fibers by old newspaper recycled fibers as polypropylene reinforcement. *Journal of Cleaner Production*, **65**, pp. 489–496.

47. Oksman, K., et al. (2016) Review of the recent developments in cellulose nanocomposite processing. *Composites Part A: Applied Science and Manufacturing*, **83**, pp. 2–18.

48. Yu, J., et al. (2011) Classical micromechanics modeling of nanocomposites with carbon nanofibers and interphase. *Journal of Composite Materials*, **45**(23), pp. 2401–2413.

49. Alcala, M., et al. (2013) All-cellulose composites from unbleached hardwood kraft pulp reinforced with nanofibrillated cellulose. *Cellulose*, **20**(6), pp. 2909–2921.

50. Álvarez-Chávez, C.R., et al. (2012) Sustainability of bio-based plastics: general comparative analysis and recommendations for improvement. *Journal of Cleaner Production*, **23**(1), pp. 47–56.

51. Kengkhetkit, N. and T. Amornsakchai (2014) A new approach to "Greening" plastic composites using pineapple leaf waste for performance and cost effectiveness. *Materials & Design*, **55**, pp. 292–299.

52. Balart, J.F., et al. (2016) Processing and characterization of high environmental efficiency composites based on PLA and hazelnut shell flour (HSF) with biobased plasticizers derived from epoxidized linseed oil (ELO). *Composites Part B: Engineering*, **86**, pp. 168–177.

53. Fazita, M.R.N., et al. (2016) Green composites made of bamboo fabric and poly (lactic) acid for packaging applications—a review. *Materials*, **9**(6).

54. Yusoff, R.B., H. Takagi, and A.N. Nakagaito (2016) Tensile and flexural properties of polylactic acid-based hybrid green composites reinforced by kenaf, bamboo and coir fibers. *Industrial Crops and Products*, **94**, pp. 562–573.

55. La Rosa, A.D., et al. (2014) Bio-based versus traditional polymer composites. A life cycle assessment perspective. *Journal of Cleaner Production*, **74**(0), pp. 135–144.

56. Lopes Silva, D.A., et al. (2014) Do wood-based panels made with agro-industrial residues provide environmentally benign alternatives?

An LCA case study of sugarcane bagasse addition to particle board manufacturing. *International Journal of Life Cycle Assessment*, **19**(10), pp. 1767–1778.

57. Joshi, S.V., et al. (2004) Are natural fiber composites environmentally superior to glass fiber reinforced composites? *Composites Part A: Applied Science and Manufacturing*, **35**(3), pp. 371–376.
58. Granda, L., et al. (2016 in press.) Fully biodegradable polylactic composites reinforced with bleached softwood fibers. *Cellulose, Chemistry and Technology*, **50**(3-4) pp. 417–422.
59. Siró, I. and D. Plackett (2010) Microfibrillated cellulose and new nanocomposite materials: a review. *Cellulose*, **17**(3), pp. 459–494.
60. Eichhorn, S.J., et al. (2010) Review: current international research into cellulose nanofibres and nanocomposites. *Journal of Materials Science*, **45**(1), pp. 1.
61. Moon, R.J., G.T. Schueneman, and J. Simonsen (2016) Overview of cellulose nanomaterials, their capabilities and applications. *Jom*, **68**(9), pp. 2383–2394.
62. Kim, J.-H., et al. (2015) Review of nanocellulose for sustainable future materials. *International Journal of Precision Engineering and Manufacturing-Green Technology*, **2**(2), pp. 197–213.
63. Mousa, M.H., Y. Dong, and I.J. Davies (2016) Recent advances in bionanocomposites: preparation, properties, and applications. *International Journal of Polymeric Materials and Polymeric Biomaterials*, **65**(5), pp. 225–254.
64. Feldman, D. (2015) Cellulose nanocomposites. *Journal of Macromolecular Science, Part A*, **52**(4), pp. 322–329.
65. Thomas, S. and L.A. Pothan, *Natural fibre reinforced polymer composites: from macro to nanoscale*. 2009: Archives contemporaines.
66. Arrieta, M.P., et al. (2015) Bionanocomposite films based on plasticized PLA–PHB/cellulose nanocrystal blends. *Carbohydrate Polymers*, **121**, pp. 265–275.
67. Wei, L., A.G. McDonald, and N.M. Stark (2015) Grafting of bacterial polyhydroxybutyrate (PHB) onto cellulose via in-situ reactive extrusion with dicumyl peroxide. *Biomacromolecules*, **16**(3), pp. 1040–1049.
68. Panaitescu, D.M., et al. (2015) Influence of thermal treatment on mechanical and morphological characteristics of polyamide 11/ cellulose nanofiber nanocomposites. *Journal of Nanomaterials*, Article ID 136204.
69. Greenberg, M.I., J. Waksman, and J. Curtis (2007) Silicosis: a review. *Disease-a-month : DM*, **53**, pp. 394–416.

70. Bledzki, A.K., H.P. Fink, and K. Specht (2004) Unidirectional hemp and flax EP- and PP-composites: influence of defined fiber treatments. *Journal of Applied Polymer Science*, **93**, pp. 2150–2156.

71. Frisoni, G., et al. (2001) Natural cellulose fibers: heterogeneous acetylation kinetics and biodegradation behavior. *Biomacromolecules*, **2**, pp. 476–482.

72. Zini, E., M. Scandola, and P. Gatenholm (2003) Heterogeneous acylation of flax fibers. Reaction kinetics and surface properties. *Biomacromolecules*, **4**, pp. 821–827.

73. Li, X., L.G. Tabil, and S. Panigrahi (2007) Chemical treatments of natural fiber for use in natural fiber-reinforced composites: a review. *Journal of Polymers and the Environment*, **15**, pp. 25–33.

74. Méndez, J.A., et al. (2007) Evaluation of the reinforcing effect of ground wood pulp in the preparation of polypropylene-based composites coupled with maleic anhydride grafted polypropylene. *Journal of Applied Polymer Science*, **105**, pp. 3588–3596.

75. Tidjani, A. (2005) Polypropylene-graft-maleic anhydride–nanocomposites: II – fire behaviour of nanocomposites produced under nitrogen and in air. *Polymer Degradation and Stability*, **87**, pp. 43–49.

76. Vilaseca, F., et al. (2007) Evaluation of the reinforcing effect of ground wood pulp in the preparation of polypropylene-based composites coupled with maleic anhydride grafted polypropylene. *Journal of Applied Polymer Science*, **105**(6), 3588–3596.

77. Rao, B.M., P.R. Rao, and B. Sreenivasulu (1999) Grafting of maleic anhydride onto polypropylene: synthesis and characterisation. *Polymer-Plastics Technology and Engineering*, **38**, pp. 967–977.

78. Lin, N., J. Huang, and A. Dufresne (2012) Preparation, properties and applications of polysaccharide nanocrystals in advanced functional nanomaterials: a review. *Nanoscale*, **4**, pp. 3274–3294.

79. Yan, L., N. Chouw, and K. Jayaraman (2014) Flax fibre and its composites – a review. *Composites Part B*, **56**, pp. 296–317.

80. Thakur, V.K. and M.K. Thakur (2014) Processing and characterization of natural cellulose fibers/thermoset polymer composites. *Carbohydrate Polymers*, **109**, pp. 102–117.

81. Lukkassen, D. and A. Meidell (2003) Advanced materials and structures and their fabrication processes. Narvik University College, Norway.

82. Hughes, M., J. Carpenter, and C. Hill (2007) Deformation and fracture behaviour of flax fibre reinforced thermosetting polymer matrix composites. *Journal of Materials Science*, **42**, pp. 2499–2511.

83. Duc, F., et al. (2014) Damping of thermoset and thermoplastic flax fibre composites. *Composites Part A: Applied Science and Manufacturing*, **64**, pp. 115–123.

84. Bos, H.L., et al. (2004) Compressive behaviour of unidirectional flax fibre reinforced composites. *Journal of Materials Science*, **39**, pp. 2159–2168.

85. Van de Weyenberg, I., et al. (2003) Influence of processing and chemical treatment of flax fibres on their composites. *Composites Science and Technology*, **63**, pp. 1241–1246.

86. Liu, W., T. Xie, and R. Qiu (2016) Improvement of properties for biobased composites from modified soybean oil and hemp fibers: dual role of diisocyanate. *Composites Part A: Applied Science and Manufacturing*, **90**, pp. 278–285.

87. Liu, W., et al. (2016) Soybean oil-based thermosets with N-vinyl-2-pyrrolidone as crosslinking agent for hemp fiber composites. *Composites Part A: Applied Science and Manufacturing*, **82**, pp. 1–7.

88. Whitacre, R., A. Amiri, and C. Ulven (2015) The effects of corn zein protein coupling agent on mechanical properties of flax fiber reinforced composites. *Industrial Crops and Products*, **77**, pp. 232–238.

89. Du, Y., et al. (2013) Pulp fiber-reinforced thermoset polymer composites: effects of the pulp fibers and polymer. *Composites Part B: Engineering*, **48**, pp. 10–17.

90. Bakare, F.O., et al. (2016) Thermomechanical properties of bio-based composites made from a lactic acid thermoset resin and flax and flax/basalt fibre reinforcements. *Composites Part A: Applied Science and Manufacturing*, **83**, pp. 176–184.

91. Moreno, M., et al. (2016) Cellulose nano-biocomposites from high oleic sunflower oil-derived thermosets. *European Polymer Journal*, **79**, pp. 109–120.

92. Menard, R., S. Caillol, and F. Allais (2016) Ferulic acid-based renewable esters and amides-containing epoxy thermosets from wheat bran and beetroot pulp: chemo-enzymatic synthesis and thermo-mechanical properties characterization. *Industrial Crops and Products*, **95**, pp. 83–95.

93. Marrot, L., et al. (2014) Multi-scale study of the adhesion between flax fibers and biobased thermoset matrices. *Materials and Design*, **62**, pp. 47–56.

94. Wu, G.M., et al. (2014) Thermoset nanocomposites from two-component waterborne polyurethanes and cellulose whiskers. *Carbohydrate Polymers*, **105**, pp. 207–213.

95. Heux, L., G. Chauve, and C. Bonini (2000) Nonflocculating and chiral-nematic self-ordering of cellulose microcrystals suspensions in nonpolar solvents. *Langmuir*, **16**, pp. 8210–8212.

96. Ortiz-Serna, P., et al. (2014) Electrical conductivity of natural rubber-cellulose II nanocomposites. *Journal of Non-Crystalline Solids*, **405**, pp. 180–187.

97. Ortiz-Serna, P., et al. (2011) Dielectric spectroscopy of natural rubber-cellulose II nanocomposites. *Journal of Non-Crystalline Solids*, **357**, pp. 598–604.

98. Martins, A.F., L.L.Y. Visconte, and R.C.R. Nunes (2002) Evaluation of natural rubber and cellulose II compositions by curing and mechanical properties. *KGK-Kautschuk und Gummi Kunststoffe*, **55**, pp. 637–641.

99. Cao, X., et al. (2013) Preparation and properties of carboxylated styrene-butadiene rubber/cellulose nanocrystals composites. *Carbohydrate Polymers*, **92**, pp. 69–76.

100. Abraham, E., et al. (2012) X-ray diffraction and biodegradation analysis of green composites of natural rubber/nanocellulose. *Polymer Degradation and Stability*, **97**, pp. 2378–2387.

101. Nagalakshmaiah, M., et al. (2015) Structural investigation of cellulose nanocrystals extracted from chili leftover and their reinforcement in cariflex-IR rubber latex. *Carbohydrate Polymers*, **136**, pp. 945–954.

102. Ladhar, A., et al. (2015) AC and DC electrical conductivity in natural rubber/nanofibrillated cellulose nanocomposites. *Journal of Molecular Liquids*, **209**, pp. 272–279.

103. Chen, Y., et al. (2015) Cellulose nanocrystals reinforced foamed nitrile rubber nanocomposites. *Carbohydrate Polymers*, **130**, pp. 149–154.

104. Cao, X., H. Dong, and C.M. Li (2007) New nanocomposite materials reinforced with cellulose nanocrystals in waterborne polyurethane. *Biomacromolecules*, **8**, pp. 899–904.

105. Mariano, M., N. El Kissi, and A. Dufresne (2016) Cellulose nanocrystal reinforced oxidized natural rubber nanocomposites. *Carbohydrate Polymers*, **137**, pp. 174–183.

106. Alhashimi, R., et al. (2014) Fabrication and characterization of polymer composites for endodontic use. *International Endodontic Journal*, **47**, pp. 574–582.

107. Iyer, K.A. and J.M. Torkelson (2015) Dispersion and property enhancements in polyolefin/soy flour biocomposites prepared via melt extrusion followed by solid-state shear pulverization. *Macromolecular Materials and Engineering*, **300**, pp. 772–784.

108. Common, A., et al. (2014) Flow and mixing efficiency characterisation in a CO_2-assisted single-screw extrusion process by residence time distribution using Raman spectroscopy. *Chemical Engineering Research and Design*, **92**, pp. 1210–1218.

109. Fenner, R.T. (1979) *Principles of Polymer Processing*. Macmillan.

110. Subramanian, M. (2015) *Basics of Polymers: Fabrication and Processing Technology*. Momentum Press.

111. Cantor, K. (2011) *Blown Film Extrusion*. Carl Hanser Verlag GmbH Co KG.

112. Franco-Marquès, E., et al. (2011) Influence of coupling agents in the preparation of polypropylene composites reinforced with recycled fibers. *Chemical Engineering Journal*, **166**, pp. 1170–1178.

113. Hassanabadi, H.M., A. Alemdar, and D. Rodrigue (2015) Polypropylene reinforced with nanocrystalline cellulose: coupling agent optimization. *Journal of Applied Polymer Science*, **132**(34).

114. Granda, L.A., et al. (2016) Semichemical fibres of *Leucaena collinsii* reinforced polypropylene: flexural characterisation, impact behaviour and water uptake properties. *Composites Part B: Engineering*, **97**, pp. 176–182.

115. Throne, J.L. (2008) *Understanding Thermoforming*, pp. 13. Cincinnati: Hanser Gardner Publications.

116. Crawford, R.J. and J.L. Throne (2001) *Rotational Molding Technology*. William Andrew.

117. Royo, J. (1989) Manual de tecnología del caucho. Consorcio nacional de industriales del caucho.

118. Vergnaud, J.-W. and J. Bouzon (1992) *Cure of Thermosetting Resins: Modelling and Experiments*, pp. 391. Springer London.

119. Rudd, C.D., et al. (1997) *Liquid Moulding Technologies: Resin Transfer Moulding, Structural Reaction Injection Moulding and Related Processing Techniques*. Elsevier.

120. Halley, P. and G. George (2009) Chemorheology of polymers: from fundamental principles to reactive processing. *Journal of Chemical Information and Modeling*, **53**, pp. 1689–1699.

121. Stritzke, B. (2009) *Custom Molding of Thermoset Elastomers. A Comprehensive Approach to Materials, Mold Design, and Processing*. Carl Hanser Verlag GmbH & Co. KG.

122. Begishev, B.P. and A.Y. Malking (1999) *Reactive Processing of Polymers*. ChemTec Publishing.

123. Juntaro, J., et al. (2012) Bacterial cellulose reinforced polyurethane-based resin nanocomposite: a study of how ethanol and processing

pressure affect physical, mechanical and dielectric properties. *Carbohydrate Polymers*, **87**, pp. 2464–2469.

124. Benhamou, K., et al. (2015) Bio-based polyurethane reinforced with cellulose nanofibers: a comprehensive investigation on the effect of interface. *Carbohydrate Polymers*, **122**, pp. 202–211.

125. Dunne, R., et al. (2016) A review of natural fibres, their sustainability and automotive applications. *Journal of Reinforced Plastics and Composites*, **35**(13), pp. 1041–1050.

126. Campos, A.R., et al. (2016) Using forest resources to develop high performance plastic compounds for the automotive industry. *International Polymer Processing*, **31**(4), pp. 423–432.

127. Galan-Marin, C., C. Rivera-Gomez, and A. Garcia-Martinez (2016) Use of natural-fiber bio-composites in construction versus traditional solutions: operational and embodied energy assessment. *Materials*, **9**(6).

128. Reixach, R., et al. (2015) Acoustic properties of agroforestry waste orange pruning fibers reinforced polypropylene composites as an alternative to laminated gypsum boards. *Construction and Building Materials*, **77**(0), pp. 124–129.

129. Korjenic, A., J. Zach, and J. Hroudova (2016) The use of insulating materials based on natural fibers in combination with plant facades in building constructions. *Energy and Buildings*, **116**, pp. 45–58.

130. Singh, B., et al., Natural fiber-based composite building materials. In *Cellulose Fibers: Bio-and Nano-Polymer Composites*. 2011, Springer. pp. 701–720.

131. Espinach, F.X., et al. (2014) Aspectos que definen la responsabilidad social del producto. *Dyna*, **81**(186), pp. 175–183.

132. Pil, L., et al. (2016) Why are designers fascinated by flax and hemp fibre composites? *Composites Part A: Applied Science and Manufacturing*, **83**, pp. 193–205.

133. Julian, F., et al. (2012) Bio-based composites from stone groundwood applied to new product development. *Bioresources*, **7**(4), pp. 5829–5842.

134. Julian, F., et al. (2012) Design and development of fully biodegradable products from starch biopolymer and corn stalk fibres. *Journal of Biobased Materials and Bioenergy*, **6**(4), pp. 410–417.

135. Tambyrajah, D. (2015) *Indulge & Explore Natural Fiber Composites. An Invitation to Product Designers*.

Chapter 4

Green Composites Reinforced with Chitin and Chitosan

Sougata Jana,[a,b] Arijit Gandhi,[c] and Kalyan Kumar Sen[a]
[a]*Department of Pharmaceutics, Gupta College of Technological Sciences, Ashram More, Asansol-713301, West Bengal, India*
[b]*Department of Health and Family Welfare, Directorate of Health Services, Government of West Bengal, Salt Lake, Kolkata-700091, West Bengal, India*
[c]*Department of Pharmaceutics, Bengal College of Pharmaceutical Sciences & Research, B.R.B Sarani, Bidhannagar, Burdwan, Durgapur-713212, West Bengal, India*
janapharmacy@rediffmail.com

4.1 Introduction

A composite material is generally composed of two or more materials in such a way that the resulting materials have certain desired properties or improved properties. Composite materials offer various advantages, such as electrical insulation, corrosion resistance, easy processability, higher stiffness and fatigue resistance, and strength. The properties of composite depend on the properties of the constituent materials. The strength and stiffness of composites are directly related to the reinforcing of fiber characteristics [1, 2].

Green Micro- and Nanocomposites
Edited by Sabu Thomas, Abitha V. K., and Hanna J. Maria

ISBN 978-981-4968-79-9 (Hardcover), 978-1-003-42756-8 (eBook)
www.jennystanford.com

Nowadays, environmental concerns result in increasing interest in natural materials. Petroleum-based materials are nonbiodegradable and toxic in nature. The biobased resins and fibers used in the green composites are biodegradable, as well as biocompatible and nontoxic. Biopolymers-based bio-fiber composites are generally considered to be called green composites. Green composites are classified into three main types on the basis of matrix and type of polymers: i) totally renewable composites, where both matrix and strengthening are from natural resources, ii) partly renewable composites, where the matrix is formed from natural resources and reinforced with a synthetic material, and iii) partly renewable composites, where the synthetic matrix is reinforced with natural polymers. Polymers used in green composites are generally natural/modified natural polymers [2, 3].

Recently, polysaccharide-based composite materials have gained attention among researchers because of their biocompatibility, renewability, and sustainability. Polysaccharide-based composites are used in the field of biomedical engineering, drug carriers, and biopolymer in film packaging. Carrageenan, alginate, chitin, chitosan (CS), starch, and cellulose-based green composites are designed by many researchers due to their mechanical properties, non-abrasive nature, strength, biodegradability, and eco-friendly nature. They are naturally and easily available and have wide economical applications compared to other synthetic biomaterials. Polysaccharide-based matrices are the most important green composites and have versatile applications in industrial and biomedical aspects [4, 5].

4.2 Chitin and Chitosan Chemical Structure and Characteristics

Chitin, poly[β-(1–4)-*N*-acetyl-D-glucosamine], is a natural material and was first identified in 1684. These biomaterials can be obtained from different sources; crabs, shrimps, cell walls, and structural membranes of fungi, green algae, yeast, and *Riftia* tubes. Chitin can be obtained by fermentation/biological methods. The extraction of chitin by fermentation is very expensive, biological extraction does not denature the chitin, is cost-effective and eco-friendly [6–8].

The chemical techniques for the extraction of chitin from sea shells involve demineralization of inorganic substances using dilute acid, deproteinization using alkaline medium, bleaching/decoloration using sodium chlorite, and to an extent deacetylation using over 40% concentration of NaOH/KOH to get chitosan [9, 10].

CS [poly-(*β*-1/4)-2-amino-2-deoxy-D-glucopyranose] is a copolymer made of D-glucosamine and *N*-acetyl-D-glucosamine bonds and *β* bonds (1–4), in which glucosamine is the predominant repeating unit in its structure [11].

CS exhibits various physicochemical properties due to its active functional groups (–NH_2 and –OH) and biological characteristics. CS solubility depends on the presence of free –NH_2 and N-acetyl groups. CS is insoluble in water, aqueous medium, concentrated acids, and common organic solvents, but it is soluble in the presence of an aqueous medium of acetic acid, hydrochloric and nitric acids, and so on. The degree of deacetylation is the rate of D-glucosamine units with respect to the total amount of *N*-acetyl-D-glucosamine that makes the CS molecule. Depending on the method of preparation, the molecular weight (MW) may vary from 300 to over 1000 kD. Generally, MW and viscosity are inversely proportional to the degree of acetylation [12–14].

CS has many attractive biomedical characteristics, such as biodegradable to normal body constituents, biocompatible, hemostatic, nonimmunogenic, and nontoxic in nature. The biological characteristics of CS are related to its MW, solubility, and degree of deacetylation [10, 15].

4.3 Preparation of Thermoset-Based Chitin and Chitosan Composites

Thermoset composites are generally made on glass, carbon/aramid fibers, usually has presence of resins such as vinyl esters, polyesters, phenolics/epoxies. Wang et al. described a slurry compounding method in which a uniform blend of chitin nanowhiskers (CW) in ethanol medium was incorporated into an epoxy matrix. The prepared CW-epoxy bio-nanocomposites were transparent and showed significant mechanical and thermal properties with tensile strength, toughness, modulus, and elongation. This favorable in toughness and strength is important for the development of nanofiller thermoset systems [16].

The natural fiber reinforced thermoset composites are based on various characteristics such as the following [17]:

1. Fibers volume fraction
2. Ratio of fiber aspect
3. Adhesion characteristics of fiber–matrix

4.4 Preparation of Thermoplastic-Based Chitin and Chitosan Composite

Thermoplastic systems are based on polymers that can be shaped in a liquid (viscous) state at temperatures higher than their glass transition temperature or higher than their melting temperature. Arce et al. introduced chitosan into the thermoplastic polyurethane (TPU) matrix through a solvent exchange method and solution casting. Thermal analyses indicated that the incorporation of chitosan into composite leads to significant changes in the glass transition temperature and melting behavior characteristic of the hard segments. Rheological analyses indicated shear-thinning behavior [18].

Recently, thermoplastic CS-based substances were successfully formulated by thermomechanical method in an internal mixer using nonvolatile (glycerol) plasticizer, volatile (water), and destructuring agent (acidic solution). TiO_2-based nanoparticles were blended into the CS-based matrix to obtain different nanocomposites [19]. The mechanical properties of thermoplastic composites can be achieved by fiber and matrix interaction.

4.5 Preparation of Rubber-Based Chitin and Chitosan Composite

Rubber-based high strength and flexible matrix can be achieved by reinforcing with CS. Zakaria et al. prepared poly(lactic acid) (PLA)-CS natural polymer/ENR (epoxidized natural rubber) composites by solution casting techniques. The characteristics of the microstructure of PLA-ENR were significantly changed by the addition of CS. The mechanical properties (tensile modulus and tensile strength) were significantly increased and the % elongation at break decreased with the incorporation of CS into the matrix [20].

Another study involved the addition of small amounts (up to 9 wt%) of chitin microsized particles to carboxylated styrene-butadiene rubber (XSBR) matrix to evaluate mechanical, thermal, and swelling behavior [21].

Kawano et al. prepared composite sheets by mixing self-assembled CNF (chitin nanofiber) with natural rubber (NR) latex stabilized by ammonia and drying under reduced pressure. Porous CNF-NR biocomposite was developed by evaporating ammonia from the CNF-NR mixture and lyophilization [22].

4.6 Preparation of Micro Chitin and Chitosan Composites

Microcomposites, as the name indicates, means composites reinforced with microsized fillers. The mechanical, electrical, thermal, optical, electrochemical, and catalytic properties of the microcomposite will differ markedly from that of the component materials. The chitin and chitosan have been utilized as the reinforcing filler to improve the properties of composites. Several methods can be used to formulate chitosan-based microcomposites.

For example, microcomposites made of biopolymers chitosan and carboxymethyl cellulose are formulated by the ionic gelation method. Here, different concentrations of two chelating agents barium chloride and sodium tripolyphosphate are used to make microcomposites [23]. In another study, chitosan–gelatin (CG) microcomposites were prepared according to the complex coacervation method and cross-linked by glutaraldehyde [24]. Jana et al. evaluated the effect of *Boswellia* gum resin on the properties of glutaraldehyde cross-linked chitosan polymer composites and their potential as oral delivery vehicles [25].

Polylactic acid, polyglycolic acid, and poly-€-caprolactone have been extensively utilized to prepare chitosan-based microcomposites. Porous polyglycolic acid/chitosan composites can be fabricated by mixing solutions in dimethyl sulfoxide and acidic water, followed by freezing, immersion in sodium hydroxide, washing with phosphate buffer, and drying [26]. Chitosan/polyglycolic acid or chitosan/poly-€-caprolactone microcomposites can be prepared by an emulsifying aqueous solution of chitosan in an organic solution of other polymer and followed by freeze-drying [27, 28]

In another work, Chameettachal et al. explored the effect of reinforcement of silk matrix with chitosan microparticles to create a viscoelastic matrix that could support the redifferentiation of expanded chondrocytes. Chitosan-reinforced scaffolds retained a statistically significant higher amount of glycosaminoglycan [29].

Several chitosan/metal ion microcomposites are also reported. For example, a new microcomposite composed of nickel ion and chitosan was prepared for the purpose of inhibiting enterovirus 71 infections [30].

4.7 Preparation of Nano Chitin and Chitosan Composites

CS-based nanocomposite materials are based on the inclusion of reinforcing fillers with nanometric scale dimensions (within 100 nm) [31]. Depending upon nature and surface property of the nanofillers, nanoparticles could exploit modifications in their properties, such as improved barrier and mechanical and higher transparency [32, 33]. Inclusion of nano-reinforcements into the CS matrix is a good technique to minimize the drawbacks of the conventional biomaterials. CS based nanocomposites are used in various sectors like pharmaceutical science, cosmetics industry, food technology, biotechnology, agriculture, textiles, etc.

Several research studies have indicated that chitosan forms nanocomposites with montmorillonite (MMT) clay, and drug-release behavior from the chitosan/MMT nanocomposite scaffolds, films, and hydrogels are extensively studied [34]. Tan et al. reported synthesis of chitosan/MMT nanocomposites in presence of hydroxylammonium cations. Here, the chitosan solution, the MMT suspension, and hydroxyaluminium oligomeric cations were mixed with a ratio of 10 mmol Al per g of MMT and stirred for 2 days at 60 °C. The mixture was finally filtered, washed, and air dried [35]. Lavorgna et al. investigated the presence of glycerol as a plasticizer in chitosan/MMT films synthesized by the solution casting techniques. The addition of glycerol as a plasticizer decreases the extent of H-bonding interactions between CS and the MMT [36]. In another study, the effect of tricetadecylmethylammonium bromide (TRIAB)-modified MMT on the mechanical characteristics of matrix composites of poly (butyl acrylate) and CS has been reported by Yu et al. The nanomatrix showed enhancement of tensile characteristic at relatively small amounts of TRIAB-modified MMT loadings [37].

The addition of nano-hydroxyapatite, bioactive glass ceramics, and the mechanical properties of composite scaffolds of chitosan can be modified which are used in tissue engineering [38]. In a research by Nikpour et al., nano-hydroxyapatite/CS composite were formulated via in situ hybridization method [39]. Chitosan/CNT nanocomposites are also produced to improve biosensor performance as well as physical and mechanical characteristics of CS materials [40].

Another approach is to prepare graphene/chitosan nanocomposite. Addition of 1 wt% graphene oxide improved the tensile strength and Young's modulus of graphene oxide-CS nanocomposites [41].

Currently, zinc oxide (ZnO) and titanium dioxide (TiO_2) are used to prepare chitosan-metal oxide composite nanoparticles that exhibit excellent photocatalytic property and these composites are stable in acidic and alkaline solvents [42].

Murugadoss and Chattopadhyay reported a new metal nanoparticle composite of Ag nanoparticles (NPs) and CS. The particle size was less than 5 nm [43].

In various conventional methods such as extrusion, injection moulding, and compression, the following parameters generally define the composite properties:

1. Specific mechanical energyShear impact
2. PressurePlasticizerTime
3. TemperatureClay dispersion

Hybrid polymer matrices containing CS and other polymers such as poly(butyl acrylate) and poly(lactic acid) are also advantageous to improve the nanocomposite properties. Cellulose is also used to prepare chitosan-based nanocomposites. For example, Tran et al.

produced a bio-composite of cellulose (CEL) and CS. The synthesized composite showed good adsorption and favorable mechanical strength [44].

4.8 Applications of Chitin/Chitosan-Based Composites

4.8.1 Tissue Engineering

The main purpose of tissue engineering is repair and regenerate of damaged tissue in animal and human body. In the field of tissue engineering, 3D (three-dimensional) porous matrixes are used for cell proliferation, adhesion, differentiation, and development of an ECM (extracellular matrix). The polymeric scaffolds contain therapeutic compounds to controlled the damage site [45]. CS-based green composites have versatile orthopedic applications [46].

For example, the poly(L-lactic acid) (PLLA)-CS hybrid matrix were fabricated for bone tissue engineering. The size and surface morphology of the hybrid matrix depend on the concentration of CS solution in PLLA matrix. The PLLA-CS matrix may be used as calcium phosphate precursors, when fabricated into calcium and phosphorus solution [47]. Gelatin is a natural derivative of collagen, which is found in our body. It has been widely used for the development of green composite with CS for promoting proliferation, cell attachment, and differentiation. Polymer and inorganic–organic composite matrix also showed functional and good mechanical properties [48].

CS-nanohydroxyapatite complex have been utilized in tissue engineering because it has good proliferative properties in osteoblasts. This matrix also showed significant bone regeneration characteristics in micro-computerized tomography study [49]. CS-based hydrogel complex has been fabricated for cartilage tissue regeneration. The most important component of cartilage regeneration is collagen type-II because it increases the adhesion and formation of clusters of chondrocytes in vitro [50, 51]. Zhao et al. investigated calcium phosphate cements (CPC)–CS–polyglactin fiber matrix composed of human umbilical cord mesenchymal stem cells (hUCMSCs) for bone tissue engineering. The CPC–CS–polyglactin scaffolds showed significant proliferation and attachment of stem cells [52]. Currently, Vyas et al. prepared CS-based scaffolds for bone tissue engineering. Macrospheric 3D scaffolds were developed by different concentration of montmorillonite and hydroxyapatite [53]. Another bio-nanocomposite composed of mesoporous silica nanoparticles (MSN)/CS was formulated by the electrospinning process. The authors stated that the MSN-CS biocomposites are promising for local implantable matrix in bone tissue engineering [54]. In another study, a matrix of CS and organoclay were prepared by the freeze-drying method for bone tissue engineering application. The prepared scaffolds showed good proliferation and biocompatibility [55].

4.8.2 Drug Delivery

Chitosan-based nanocomposites are important for the delivery of drug to the target organ. It has numerous properties such as biocompatibility, biodegradability, non-toxicity, and good adhesion sorption.

CS is an important material for polymer blends/composites. Using different fabrication technology, researchers may obtain films, gels, matrix/network structure, and beads of different sizes and surface morphology [56]. In a study, CS and sodium alginate (Na-ALG) with different concentration were mixed with various wt% of Cloisite 30B(C-30B) solution by solvent evaporation techniques. Cloisite 30B was used as co-emulsifier in the fabrication of nanocomposite. Different concentrations of curcumin were loaded in the matrix of CS-ALG/ C- 30B. The drug release kinetics study indicated swelling controlled, non-Fickian type pattern [57]. Regiel-Futyra et al. described that the nanocomplex showed significant antibacterial activity. The CS was used as reducing and stabilizing agent for the development of gold nanoparticles. The fabricated formulation showed antibacterial activity against strains of *Staphylococcus aureus* and *Pseudomonas aeruginosa.* The formulated nanocomposites were observed to be nontoxic against mammalian somatic and tumoral cells [58].

Superporous hydrogel particles containing metoprolol succinate were prepared by Chavda et al. for intestinal delivery. Chitosan and

HPMC (hydroxypropyl methylcellulose) were used as bioadhesive and drug release retardant polymers. The fabricated hydrogel showed remains of bioadhesive up to 8 h in intestine and the drug release followed Hixson–Crowell model with anomalous non-Fickian type [59].

Very recently, Rishabha et al. prepared the composites of chitosan and guar gum for the fast-disintegrating drug delivery systems by mixing both polymers in same proportion at regulated temperature. Six batches of tablets were prepared with composites in different ratios using phenytoin sodium as model drug. The optimized formulation showed best in vitro release and released about 90% drug within the first five minutes [60]. In another study, water-insoluble CS-based composite films were developed by hot press techniques [61]. The authors concluded that the polysaccharide composite is promising for sustained drug-release characteristics in physiological environment. Hu et al. synthesized alginate and carboxymethyl chitosan (CMCS) pH-sensitive beads for drug delivery by ionic cross-linked method. BSA loaded CS-Alg beads showed pH-sensitive sustained release for oral drug delivery system [62].

Zinc-pectin-CS micro-composite were formulated as colon-specific carrier by Das et al. for colon delivery. Model drug (resveratrol) was used for the treatment of colon diseases. The prepared micro-composite composed of pectin and drug ratio (3:1) with 1% CS, with 120 min cross-linking time showed colon-targeting drug release [63].

The amoxicillin loaded CS-PVA as films cross-linked by GA (glutaraldehyde) indicated pH responsive systems, investigated by Altinisik and Yurdakoc [64]. Wang et al. have observed PVA and CMCS blend for local drug delivery applications [65].

Inter-polymeric nanocomposite of CS-egg albumin were fabricated by Jana et al. (2013). Further nanocomplex was stabilized by PEG 400 [66]. In another study, Jana et al. (2014) synthesized CS-egg albumin nanocomposite by ionic crosslinking and coagulation method by heat. Nanocomposite was containing aceclofenac for oral drug delivery. Prepared nanoparticles were characterized by the DSC, FE-SEM, P-XRD, and FTIR methods. The best formulation average particle size was 352.90 nm (Fig. 4.1a) and surface morphology were observed by FE-SEM analysis (Fig. 4.1b) [67].

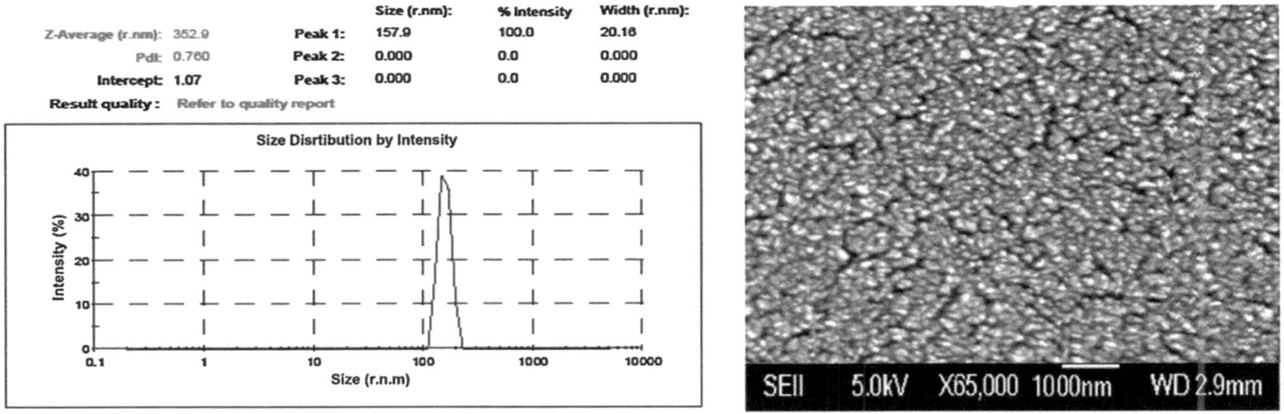

(a) (b)

Figure 4.1 (a) Particle size distribution of Na-TPP (2% w/v) aceclofenac-loaded CS-egg albumin nanocomposite (F-6) (b) FE-SEM image of Na-TPP (2% w/v) aceclofenac-loaded CS-egg albumin nanocomposite (F-6) at 65,000× magnification (with permission form Elsevier).

In current year, Jana et al. (2017) fabricated IPN nanocomposites of CS and locust bean gum (LBG) by using glutaraldehyde (GA) as cross-linker (Fig. 4.2). FTIR spectroscopy analysis showed the chemical compatibility between drug and polymers. The fabricated nanocomposites were promising for oral drug delivery application [68].

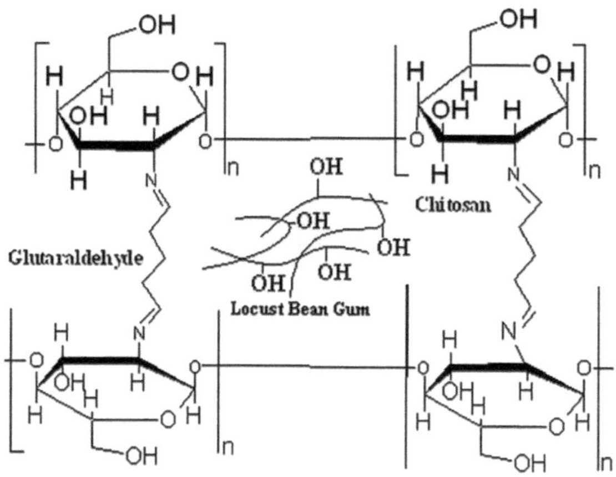

Figure 4.2 The reaction mechanism of GA crosslink CS-LBG IPN (with permission from Elsevier [68]).

4.8.3 Wound Healing

CS is globally used as a wound-dressing biomaterial for stimulating cell adhesion and proliferation. Researchers observed that CS-based membrane showed significant wound-healing effects by increasing the re-epithelialization and granular layer [69, 70]. CS-collagen systems showed significant wound-dressing characteristics due to their biocompatibility and antibacterial and liquid-absorbing properties [71]. Chitosan and cellulose composites showed anti-inflammatory activity through the reduction of tumor necrosis factor-*a* (TNF-*a*) and interleukin-6 (IL-6). The prepared composites revealed to inhibit the growth of both microorganisms (gram positive and negative) [72].

CS-PEGDA (polyethylene glycol diacrylate) blend films were synthesized by Michael addition reaction for wound-dressing purposes. The mechanical/swelling characteristics of CS were enhanced after the chemical modification. SEM results showed that the microporous surface structure of the CS-PEGDA films was good to grow and proliferate. CS-PEGDA films were promising as wound-dressing material [73]. In another research study, Menda et al. proved that a blend of CS and plant extract have been potential for wound-healing properties. Authors developed two composite films formulations: CS–aloe vera extract and CS–*C. officinalis* petal extract for wound healing purpose. CS-Aloe Vera extract films showed higher antimicrobial activity and water absorption capacity in compared to the formulation of CS-*C. officinalis* petal extract. But higher tensile strength observed in CS-based formulation containing *C. officinalis* petal extract. Authors also stated that CS-based film containing 25% aloe vera is significant as wound-dressing materials [74].

Curcumin-loaded CS-gelatin composite sponge was prepared by Nguyen et al. These CS-gelatin composite sponge are important for wound healing purpose [75]. Kouchak et al. prepared CS-PVA composite films casting/solvent casting evaporating technique. The prepared films indicated higher activity against *Pseudomonas aeruginosa* than drug-free CS films. The presence of PVA improves many characteristics of CS-nitrofurazone films and is useful as burn wounds dressing material [76]. Recently, Zou et al. developed tourmaline/chitosan (Tur/CS) composite films for wound healing applications. When the composite films were transplanted on porcine full-thickness burn wounds, histological results demonstrated that the Tur/CS group with 1/40 mass ratio had a significantly formed higher number blood vessels and regeneration of dermis [77]. Another study described fibrin–CS–sodium alginate complex for wound dressing material. SEM result showed complex nature of sheet developed successfully [78].

4.8.4 Protein and Gene Delivery

CS MPs and NPs are important for the delivery of protein due to their mucoadhesive properties. Dastan and Turan prepared CS–DNA MPs and showed a sustained-release profile of DNA from the prepared MPs, with a potential transfer of the DNA into human embryonic kidney and HeLa cell lines, Swiss 3T3 [79]. Insulin delivery through nasal route using CS MPs was demonstrated by Varshosaz et al. who reported that the insulin-loaded MPs showed a 67% lowering in the blood glucose level compared to insulin administered by the intravenous route (with absolute insulin bioavailability of 44%) [80]. Wang et al. fabricated quaternized CS that was intercalated into the interlayers of rectorite to obtain a new polymer/layered silicate nanocomposite. In vitro and in vivo toxicity studies revealed that the NPs were biocompatible and non-toxic [81]. Min et al. showed that, among CS of different molecular weights, the 45 kD form exhibited significantly higher gene delivery efficiency than the 15 kD and 100 kD forms. Using the 45 kDa form of CS, they formulated composite polyethyleneimine/DNA/CS (PDCS) complexes consisting of the core part where DNA chains were tightly condensed by polyethyleneimine and the outer layer lined with CS and found that an optimal PDC complex exhibits a significantly higher transfection efficiency in a high serum condition and a lower toxicity than monolithic polyethyleneimine/DNA (PD) complex [82]. In another research work, PLGA/CS composite foams loaded with luciferase plasmid were fabricated by a combination of spray drying and supercritical CO_2 foaming techniques. The resultant composite foams showed good morphology and CS was found to be poorly crystallized in the PLGA matrixes. The composite foams exhibited a sustained release of DNA (5–9 weeks) with decreasing release rate upon increasing content of chitosan [83].

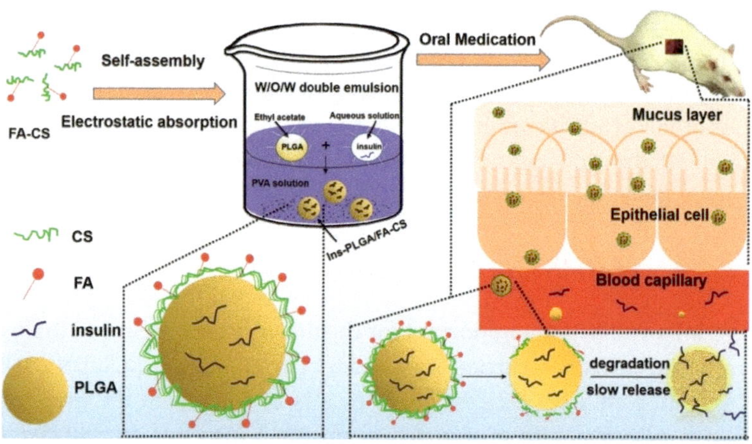

Figure 4.3 The schematic representation for preparation of Ins-PLGA/FA-CS nanocarriers (with permission from Elsevier [85]).

Eroglu et al. developed a unique nanomaterial, poly(2-hydroxyethyl methacrylate) CS NPs, which consisted of poly(2-hydroxyethyl methacrylate) NPs surrounded by a CS cationic shell, and we used this for encapsulation of a respiratory syncytial virus (RSV)-F gene construct (a model for a DNA vaccine). The authors clearly stated that the positively charged formulation were able to bind to the negatively charged cell membrane and were taken up by endocytosis, in Cos-7 cells. [84]. In current year, Xu et al. (2017) formulated folic acid modified CS (FA-CS) and poly(lactide-co-glycoside) (PLGA) nanocomposites (PLGA/FA-CS) by electrostatic self-assembly techniques (Fig. 4.3) for oral delivery of insulin [85].

Particle size and surface characteristics were evaluated by the SEM, DLS, and TEM analysis (Fig. 4.4). Stability of insulin loaded PLGA/FA-CS nanocomposites were evaluated in 10% fetal bovine serum (FBS) solution at 37 °C against incubation time and result showed that no significant change in the size of nanocomposite after 24 h (Figure 4.4D).

4.8.5 Dentistry

CS properties like biocompatibility, anti-inflammatory, and others could give promising results in periodontal care or wound healing after teeth extractions. Chitosan in the form of gel and hydrogels applies to the treatment of chronic periodontitis and canker sores.

Ali et al. showed antibacterial activity and hydrolytic stability of RDCRM (resin dental composite restorative materials) composed of CS. The antibacterial property of micro-hybrid and flowable RDCRM containing 0, 0.25, 0.5, and 1% wt/wt CS against *Lactobacilli casei* bacteria was investigated using agar diffusion test and direct contact methods and antimicrobial activity may prevent the problem of secondary caries [86].

Thiolated CS-based formulations are also used in mucoadhesive patches to cure dental caries. The sustained release (SR) of antibacterial medicament inhibits the growth of cariogenic *Streptococcus mutans* [87]. Recently, Ruan et al. employed a CS-based hydrogel as a delivery medium for amelogenin with the aim of rejuvenating the aligned crystal structure. The use of CS imparts a dual effect of offering a protective effect against secondary caries corresponding to its antibacterial properties along with not influencing enamel crystal orientation [88]. Antioxidant CS

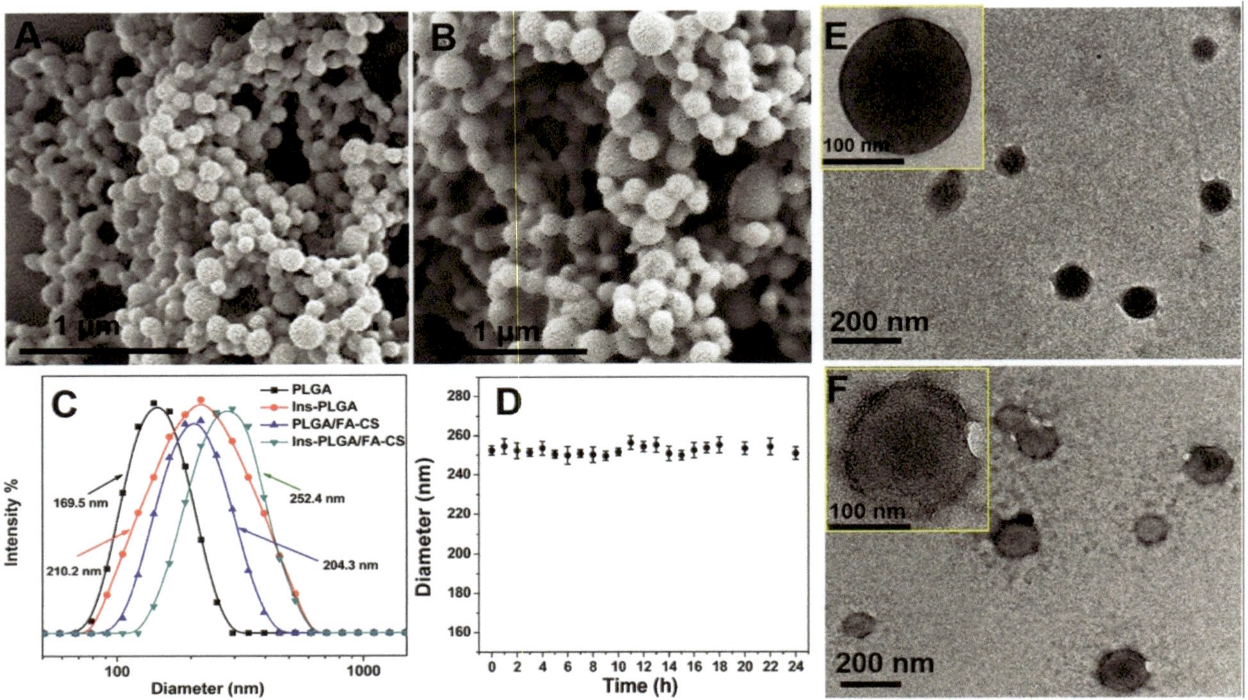

Figure 4.4 SEM image of PLGA/FA-CS (A) and Ins-PLGA/FA-CS nanocarriers (B), DLS distribution of PLGA, Ins-PLGA, PLGA/FA-CS, and Ins-PLGA/FA-CS in water (C), stability of insulin-loaded PLGA/FA-CS nanocarriers in 10% fetal bovine serum (FBS) solution at 37 °C against incubation time (D), TEM images of PLGA (E), and PLGA/FA-CS nanoparticles (F) (with permission from Elsevier [85]).

hydrogels with propolis, β-carotene, and nystatin were evaluated and translated significant grounds toward delivering robust dentine bonding systems with a concomitant increase in shear bond strength [89]. Yang et al. investigated the use of dental pulp stem cells cultured on a collagen–CS complex and were also able to form a dentine–pulp complex [90]. Bonifacio et al. applied a poly(acrylic acid) (PAA) layer by electropolymerization followed by electrochemical deposition of a CS-gallium layer on a titanium surface to reduce orthopedic and/or dental implants failure [91].

Antibacterial Activity of a chitosan-polyvinyl alcohol-Ag^+-tobermorite composite for periodontal repair was evaluated by Hurt et al. The alkaline dissolution products of the tobermorite lattice buffered the acidic breakdown products of the chitosan polymer and the presence of silver ions resulted in marked antimicrobial action against *S. aureus*, *P. aeruginosa*, and *E. coli* [92].

4.8.6 Food Packaging Applications

Various biopolymers have been employed to develop biodegradable food packaging materials. But the usage of biopolymers has been limited due to the poor mechanical and barrier properties. These properties can be enhanced by adding reinforcing compounds or fillers to form composites. Blends of chitosan and anionic polymers have been reported to have improved mechanical and barrier properties when compared to those made of chitosan only. Improvements in mechanical properties, better performance in terms of water vapor permeability, and lower water solubility have been reported for combinations of chitosan with other polysaccharides, such as starch, pectin or alginate, compared to chitosan membranes [94, 95].

For example, Al-Naamani et al. prepared chitosan–zinc oxide (ZnO) nanocomposite coatings on polyethylene (PE) films. [96]. PE packaging films were coated with chitosan in order to introduce the antibacterial activity to the films. In a study conducted by Theapsak et al., the surfaces of the PE films were modified by dielectric barrier discharge (DBD) plasma before chitosan coating [97].

In another work, low-density polyethylene (LDPE)/CS composites were tested for minced poultry meat packaging. It has been established that the composites containing 3–6 wt% CS showed superior properties, compared with commercially available foils, when applied to meat products. It was found that the number of germs decreased for meat packed in the obtained compositions by 104 to 105 times [98]. Pereda et al. designed composite films as potentially edible food packaging. These films were prepared by casting film-forming emulsions based on chitosan/glycerol/olive oil containing dispersed cellulose nanocrystals (CNs) [99].

4.8.7 Textile Industry

The large surface area and moisture retaining property of textile structures support microorganisms' growth. Over the last few years, several researches have been conducted in order to minimize microorganisms' growth on textiles. Therefore, to impart antimicrobial ability to textiles, several approaches have been studied, being mainly divided into the inclusion of antimicrobial agents in the textile polymeric fibers or their grafting onto the polymer surface.

Buşilă et al. synthesized antimicrobial textile finishing based on Ag:ZnO nanoparticles/chitosan biocomposites [100]. Rajendran et al. developed herbal chitosan nanocomposite for protective clothing [101]. In another study, chitosan–ZnO composites were prepared by hydrothermal method by Rajendranand and Sivalingam. Antibacterial activity of chitosan–ZnO composite is shown to be higher for gram-positive bacterium compared to gram-negative bacterium. The authors stated that the preparation of chitosan–ZnO composites by this way will give bulk quantity with low cost for industrial bio-finishing [102]. Wang et al. developed chitosan-metal complexes for antimicrobial activity against two gram-positive bacteria (*S. aureus* and *S. epidermidis*), two gram-negative bacteria (*E. coli* and *P. aeruginosa*), and two fungi (*Candida albicans* and *Candida parapsilosis*). The authors concluded that the phenomenon was due to the stronger positive charge after complexation [103].

Applications of chitin/chitosan-based composite materials are given in Table 4.1.

Table 4.1 Application summary of chitin/chitosan-based composite materials

Composite	Purpose	References
Chitin-graphene oxide–ZnO	Sensing, catalysis, and energy storage applications.	104
Bioactive glass-chitosan-alginate	Increase of the mechanical properties	105
Silver nanoparticles together with chitosan	Homogenous nanofibers structures with a high efficiency against bacterial growth	106
Chitosan-hydroxyapatite	Reconstructing bone defects	107
Nano-hydroxyapatite-chitosan-caboxymethyl cellulose	Bone tissue engineering	108
Chitosan-calcium phosphate	Formed a unique core–shell structure with better mechanical properties	109
Chitin nanofibrils-chitosan glycolate	Wound medicaments	110
Chitosan-sodium alginate	Study of drug-release kinetics and mechanism	57
Zinc-pectin-chitosan	Colon-specific drug release	63
PLGA/chitosan	Sustained release of DNA	83
Chitosan-cellulose acetate	Waste water treatment	93
Chitosan-guar gum	Fast disintegrating drug delivery	60

4.9 Conclusions

Composites, obtained by adding fillers to biopolymers like chitin or chitosan result in very promising materials since they show improved properties with preservation of the material biodegradability without toxicity. Hybrid polymer matrices containing chitosan and other biodegradable polymers are suggested to be advantageous in improving the composite properties. Several research studies suggest a range of applications of composites in which chitosan can be moulded into a variety of structures such as hydrogels, films, powders, and particles. Various composites like metal-matrix composites, nanocomposites, fiber-reinforced composites, and hybrid composites are commonly used in various industries. The development of novel polymeric materials based on chitosan presents a very interesting and promising approach to explore the application of chitosan-based composites. The development of optimum formulations and processing techniques to obtain the desired properties, as well as cost reduction of the composites, still needs attention.

References

1. Sahari, J. and Sapuan, S.M. (2011). Natural fibre reinforced biodegradable polymer composites, *Reviews on Advanced Materials Science*, **30**, pp. 166–174.
2. La Mantia, F.P. and Morreale, M. (2011). Green composites: A brief review, *Composites Part A: Applied Science and Manufacturing*, **42**, pp. 579–588.
3. Kandpal, B.C., Chaurasia, R. and Khurana, V. (2015). Recent advances in green composites – A review, *International Journal for Technological Research in Engineering*, **2**, pp. 742–747.
4. Zini, E. and Scandola, M. (2011). Green composites: An overview, *Polymer Composites*, **32**, pp. 1905–1915.
5. Cheung, H.Y., Ho, M.P., Lau, K.T., Cardona, F. and Hui, D. (2009). Natural fibre-reinforced composites for bioengineering and environmental engineering applications, *Composites Part B: Engineering*, **40**, pp. 655–663.
6. Das, S. and Ganesh, E.A. (2010). Extraction of chitin from trash crabs (Podophthalmus vigil) by an eccentric method, *Journal of Biological Sciences*, **2**, pp. 72–75.
7. Cauchie, H.M. (2002). Chitin production by arthropods in the hydrosphere, *Hydrobiologia*, **470**, pp. 63–96.
8. Ifuku, S., Nogi, M., Yoshioka, M., Morimoto, M., Yano, H. and Saimoto, H. (2010). Fibrillation of dried chitin into 10-20 nm nanofibers by a simple grinding method under acidic conditions, *Carbohydrate Polymers*, **81**, pp. 134–139.
9. Jana, S., Gandhi, A., Sen, K.K., and Basu, S.K. (2013). *Biomedical Applications of Chitin and ChitosanDerivatives*, Ed. Kim, S-K., Chapter 18 "Chitin and Chitosan Derivatives: Advances in Drug Discovery and Developments" (Taylor & Francis Group, LLC, USA), pp. 337–360.

10. Jana, S., Sen, K.K., Gandhi, A., Jana, S., and Roy, C. (2017). *Drug Delivery Applications of Chitosan*, Ed. Sudha, P.N., Chapter 12 "Industrial Applications of Marine Biopolymers" (CRC Press, Boca Raton, USA), pp. 305–325.

11. Jana, S., Gandhi, A., Sen, K.K., and Basu, S.K. (2011). Natural Polymers and their application in drug delivery and biomedical field, *Journal of PharmaSciTech*, **1**, pp. 16–27.

12. Rinaudo, M. (2006). Chitin and chitosan: properties and applications, *Progress in Polymer Science*, **31**, pp. 603–632.

13. Kim, I.Y., Seo, S.J., and Moon, H.S. (2008). Chitosan and its derivatives for tissue engineering applications, *Biotechnology Advances*, **26**, pp. 1–21.

14. Jana, S., Sen, K.K., Gandhi, A., Jana, S. (2016). *Polysaccharides from Marine Sources and Their Pharmaceutical Approaches*, Ed. Kim, S-K., Chapter 15 "Marine Glycobiology: Principles and Applications" (Taylor & Francis Group, LLC, USA), pp. 209–225.

15. Stefan, J., Lorkowska-Zawicka, B., Kaminski, K., Szczubialka, K., Nowakowska, M. and Korbut, R. (2014). The current view on biological potency of cationically modified chitosan, *Journal of Physiology and Pharmacology*, **65**, pp. 341–347.

16. Wang, M., Xue, H., Feng, Z., Cheng, B. and Yang, H. (2017). Increase of tensile strength and toughness of bio-based diglycidyl ether of bisphenol A with chitin nanowhiskers, *PLoS One*, **12**, e0177673.

17. Mitra, B.C. (2014). Environment friendly composite materials: Biocomposites and green composites, *Defence Science Journal*, **64**, pp. 244–261.

18. Arce, F., Pedrazzoli, D. and Manas-Zloczower, I. (2016). Thermoplastic polyurethane chitosan/cellulose nanocrystals composites for wound healing applications. *SPE ANTEC™ Indianapolis*, **2016**, pp. 1757–1761.

19. Cano, L., Pollet, E., Avérous, L. and Tercjak, A. (2017). Effect of TiO_2 nanoparticles on the properties of thermoplastic chitosan-based nano-biocomposites obtained by mechanical kneading, *Composites Part A: Applied Science and Manufacturing*, **93**, pp. 33–40.

20. Zakaria, Z., Islam, M.S., Hassan, A., Haafiz, M.K.M., Arjmandi, R., Inuwa, M. and Hasan, M. (2013). Mechanical properties and morphological characterization of PLA/chitosan/epoxidized natural rubber composites, *Advances in Materials Science and Engineering*, **2013**, pp. 1–7.

21. Visakh, P.M., Monti, M., Puglia, D., Rallini, M., Santulli, C., Sarasini, F., Thomas, S. and Kenny, J.M. (2012). Mechanical and thermal properties of crab chitin reinforced carboxylated SBR composites, *eXPRESS Polymer Letters*, **6**, pp. 396–409.

22. Kawano, A., Yamamoto, K. and Kadokawa, J-I. (2017). Preparation of self-assembled chitin nanofiber-natural rubber composite sheets and porous materials, *Biomolecules*, **7**, pp. 1–10.

23. Samrot, A.V., Akanksha, Jahnavi, T., Padmanaban, S., Philip, S-A., Burman, U. and Rabel, A.M. (2016). Chelators influenced synthesis of chitosan–carboxymethyl cellulose microparticles for controlled drug delivery, *Applied Nanoscience*, **6**, pp. 1219–1231.

24. Konuklu, Y. and Paksoy. H.O. (2015). The preparation and characterization of chitosan–gelatin microcapsules and microcomposites with fatty acids as thermal energy storage materials, *Energy Technology*, **3**, pp. 503–508.

25. Jana, S., Laha, B. and Maiti, S. (2015). *Boswellia* gum resin/chitosan polymer composites: Controlled delivery vehicles for aceclofenac. *International Journal of Biological Macromolecules*, **77**, pp. 303–306.

26. Wang, Y.C., Lin, M.C., Wang, D.M. and Hsieh, H.J. (2003). Fabrication of a novel porous PGA-chitosan hybrid matrix for tissue engineering, *Biomaterials*, **24**, pp. 1047–1057.

27. Lee, J.Y., Nam, S.H., Im, S.Y., Park, Y.J., Lee, Y.M., Seol, Y.J., Chung, C.P. and Lee, S.J. (2002). Enhanced bone formation by controlled growth factor delivery from chitosan-based biomaterials, *Journal of Controlled Release*, **78**, pp. 187–197.

28. Im, S.Y., Cho, S.H., Hwang, J.H. and Lee, S.J. (2003). Growth factor releasing porous poly (ε-caprolactone)-chitosan matrices for enhanced bone regenerative therapy, *Archives of Pharmacal Research*, **26**, pp. 76–82.

29. Chameettachal, S., Murab, S., Vaid, R., Midha, S. and Ghosh, S. (2017). Effect of visco-elastic silk-chitosan microcomposite scaffolds on matrix deposition and biomechanical functionality for cartilage tissue engineering, *Journal of Tissue Engineering and Regenerative Medicine*, **11**, pp. 1212–1229.

30. Lin, Y-C. and Chang, C-H. (2014). In vitro inhibition of enterovirus 71 infection with a nickel ion/chitosan microcomposite, *Virus Research*, **190**, pp. 17–24.

31. Averous, L. and Boquillon, N. (2004). Biocomposites based on plasticized starch: Thermal and mechanical behaviours, *Carbohydrate Polymers*, **56**, pp. 111–122.

32. Picard, E., Vermogen, A., Gerard, J.F. and Espuche, E. (2007). Barrier properties of nanocomposite membranes: Influence of clay content and dispersion state consequences on modeling, *Journal of Membrane Science*, **292**, pp. 133–144.

33. Sinha Ray, S. and Okamoto, M. (2003). Polymer/layered silicate nanocomposites: A review from preparation to processing, *Progress in Polymer Science*, **28**, pp. 1539–1641.

34. Wang, X.Y., Liu, B., Sun, R.C. and Wu, J. (2010). *Huanan Ligong Daxue Xuebao/J. South China Univ. Technol. (Nat. Sci.)*, **38**, pp. 96–101.

35. Tan, W., Zhang, Y.H., Szeto, Y.S. and Liao, L.B. (2007). Preparation and characterization of chitosan/hdroxy-aluminum pillared montmorillonite nanocomposites, *Key Engineering Materials*, **334/335**, pp. 825–828.

36. Lavorgna, M., Piscitelli, F., Mangiacapra, P. and Buonocore, G.G. (2010). Study of the combined effect of both clay and glycerol plasticizer on the properties of chitosan films, *Carbohydrate Polymers*, **82**, pp. 291–298.

37. Yu, L., Li, L., Wei'an, Z. and Yue'e, F. (2004). A new hybrid nanocomposite prepared by graft copolymerization of butyl acrylate onto chitosan in the presence of organophilic montmorillonite, *Radiation Physics and Chemistry*, **69**, pp. 467–471.

38. Peter, M., Binulal, N.S., Soumya, S., Nair, S.V., Furuike, T., Tamura, H. and Jayakumar, R. (2010). Nanocomposite scaffolds of bioactive glass ceramic nanoparticles disseminated chitosan matrix for tissue engineering applications, *Carbohydrate Polymers*, **79**, pp. 284–289.

39. Nikpour, M.R., Rabiee, S.M. and Jahanshahi, M. (2012). Synthesis and characterization of hydroxyapatite/chitosan nanocomposite materials for medical engineering applications, *Composites: Part B*, **43**, pp. 1881–1886.

40. Luo, X.M., Morrin, A., Killard, A.J. and Smyth, M.R. (2006). Application of nanoparticles in electrochemical sensors and biosensors, *Electroanalysis*, **16**, pp. 319–326.

41. Kang, X., Wang, J., Wu, H., Aksay, I.A., Liu, J. and Y. Lin. (2009). Glucose oxidase-graphene-chitosan modified electrode for direct electrochemistry and glucose sensing, *Biosensors and Bioelectronics*, 25, pp. 901–905.

42. Zhu, Y., Cao, H., Tang, L., Yang, X. and Li, C. (2009). Immobilization of horseradish peroxidase in three-dimensional macroporous TiO_2 matrices for biosensor applications, *Electrochimica Acta*, **54**, pp. 2823–2827.

43. Murugadoss, A. and Chattopadhyay, A. (2008). A 'green' chitosan-silver nanoparticle composite as a heterogeneous as well as micro-heterogeneous catalyst, *Nanotechnology*, **19**, 015603.
44. Tran, C.D., Duri, S., Delneri, A. and Franko, M. (2013). Chitosan-cellulose composite materials: Preparation, characterization and application for removal of microcystin, *Journal of Hazardous Materials*, **15**, pp. 355–366.
45. Prabaharan, M., Rodriguez-Perez, M.A., de Saja, J.A. and Mano, J.F. (2007). Preparation and characterization of poly(L-lactic acid)-chitosan hybrid scaffolds with drug release capability, *Journal of Biomedical Materials Research Part B: Applied Biomaterials*, **81**, pp. 427–434.
46. Di Martino, A., Sittinger, M. and Risbud, M.V. (2005). Chitosan: A versatile biopolymer for orthopaedic tissue-engineering, *Biomaterials*, **26**, pp. 5983–5990.
47. Mano, J.F., Hungerford, G. and Gómez Ribelles, J.L. (2008). Bioactive poly(L-lactic acid)-chitosan hybrid scaffolds, *Materials Science and Engineering: C*, **28**, pp. 1356–1365.
48. Sundaram, M.N., Deepthi, S. and Jayakumar, R. (2015). *Chitosan-Gelatin Composite Scaffolds in Bone Tissue Engineering*, Ed. Dutta, P.K., Chapter 5 "Chitin and Chitosan for Regenerative Medicine" (Springer, New Delhi), pp. 99–121.
49. Lee, J.S. and Venkatesan, S.D. (2014). In vivo study of chitosan-natural nano hydroxyapatite scaffolds for bone tissue regeneration, *International Journal of Biological Macromolecules*, **67**, pp. 360–366.
50. Choi, B., Kim, S. and Lin, B. (2015). Visible-light-initiated hydrogels preserving cartilage extracellular signaling for inducing chondrogenesis of mesenchymal stem cells, *Acta Biomaterialia*, **12**, pp. 30–41.
51. Zhu, Y., Wu, H., Sun, S., Zhou, T., Wu, J. and Wan, Y. (2014). Designed composites for mimicking compressive mechanical properties of articular cartilage matrix, *Journal of the Mechanical Behavior of Biomedical Materials*, **36**, pp. 32–46.
52. Zhao, L., Burguera, E.F., Xu, H.H.K., Amin, N., Ryou, H. and Arola, D.D. (2010). Fatigue and human umbilical cord stem cell seeding characteristics of calcium phosphate–chitosan–biodegradable fiber scaffolds, *Biomaterials*, **31**, pp. 840–847.
53. Vyas, V., Kaur, T. and Thirugnanam, A. (2017). Chitosan composite three dimensional macrospheric scaffolds for bone tissue engineering, *International Journal of Biological Macromolecules*, **104**, 1946–1954.

54. Li, K., Sun, H., Sui, H., Zhang, Y., Liang, H., Wu, X. and Zhao, Q. (2015). Composite mesoporous silica nanoparticle/chitosan nanofibers for bone tissue engineering, *RSC Advances*, **5**, pp. 17541–17549.

55. Demir, A.K. (2017). Evaluation of chitosan/organoclay composite as bone tissue engineering scaffold, *Hittite Journal of Science and Engineering*, **4**, pp. 45–49.

56. Sahoo, D., Sahoo, S., Mohanty, P., Sasmal, S. and Nayak, P.L. (2009). Chitosan: A new versatile biopolymer for various applications, *Designed Monomers and Polymers*, **12**, pp. 377–404.

57. Malesu, V.K., Sahoo, D. and Nayak, P.L. (2011). Chitosan–sodium alginate nanocomposites blended with Cloisite 30B as a novel drug delivery, *International Journal of Applied Biology and Pharmaceutica Technology*, **2**, pp. 402–411.

58. Regiel-Futyra, A., Kus-Likiewicz, M., Sebastian, V., Irusta, S., Arruebo, M., Stochel, G. and Kyzio, A. (2015). Development of non-cytotoxic chitosan-gold nanocomposites as efficient antibacterial materials, *ACS Applied Materials & Interfaces*, **7**, pp. 1087–1099.

59. Chavda, H., Modhia, I., Mehta, A., Patel, R. and Patel, C. (2013). Development of bioadhesive chitosan superporous hydrogel composite particles based intestinal drug delivery system, *BioMed Research International*, **2013**, Article ID 563651, pp. 1–10.

60. Rishabha, M., Pramod, K.S. and Susheel, K.D. (2017). Preparation and application of novel chitosan-guar gum composite as drug delivery carrier, *Novel Approaches in Drug Designing & Development*, **2**, pp. 1–7.

61. Hashizume, M., Murata, Y., Iijima, K. and Shibata, T. (2016). Drug loading and release behaviors of freestanding polysaccharide composite films, *Polymer Journal*, **48**, pp. 545–550.

62. Hu, Y., Peng, J., Ke, L., Zhao, D., Zhao, H., and Xiao, X. (2016). Alginate/ carboxymethyl chitosan composite gel beads for oral drug delivery, *Journal of Polymer Research*, **23**, 129.

63. Das, S., Chaudhury, A. and Ng, K.Y. (2011). Preparation and evaluation of zinc-pectin-chitosan composite particles for drug delivery to the colon: Role of chitosan in modifying in vitro and in vivo drug release, *International Journal of Pharmaceutics*, **15**, pp. 11–20.

64. Altinisik, A. and Yurdakoc, K. (2014). Chitosan/poly(vinyl alcohol) hydrogels for amoxicillin release, *Polymer Bulletin*, **71**, pp. 759–774.

65. Wang, L-C., Chen, X-G., Zhong, D-Y. and Xu, Q-C. (2007). Study on poly(vinyl alcohol)/carboxymethyl-chitosan blend film as local drug delivery system, *Journal of Material Science*, **18**, pp. 1125–1133.

66. Jana, S., Maji, N., Nayak, A.K., Sen, K.K. and Basu, S.K. (2013). Development of chitosan-based nanoparticles through inter-polymeric complexation for oral drug delivery, *Carbohydrate Polymers*, **98**, pp. 870–876.

67. Jana, S., Manna, S., Nayak, A.K., Sen, K.K. and Basu, S.K. (2014). Carbopol gel containing chitosan-egg albumin nanoparticles fortransdermal aceclofenac delivery, *Colloids and Surfaces B: Biointerfaces*, **114**, pp. 36–44.

68. Jana, S., and Sen, K.K. (2017). Chitosan - Locust bean gum interpenetrating polymeric network nanocomposites for delivery of aceclofenac, *International Journal of Biological Macromolecules*, **102**, pp. 878–884.

69. Wang, B., Chen, K., and Jiang, S. (2006). Chitosan-mediated synthesis of gold nanoparticles on patterned poly(dimethylsiloxane) surfaces, *Biomacromolecules*, **7**, pp. 1203–1209.

70. Yi, H., Wu, L.-Q. and Bentley, W.E. (2005). Biofabrication with chitosan, *Biomacromolecules*, **6**, pp. 2881–2894.

71. Wang, C.C., Su, C.H. and Chen, C.C. (2008). Water absorbing and antibacterial properties of N-isopropyl acrylamide grafted and collagen/chitosan immobilized polypropylene nonwoven fabric and its application on wound healing enhancement, *Journal of Biomedical Materials Research*, **84A**, pp. 1006–1017.

72. Harkins, A.L., Duri, S., Kloth, L.C. and Tran, C.D. (2013). Chitosan–cellulose composite for wound dressing material. Part 2. Antimicrobial activity, blood absorption ability, and biocompatibility, *Journal of Biomedical Materials Research B: Applied Biomaterials*, **102**(6), 1199–1206.

73. Zhang, Z., Yang, D. and Nie, J. (2008). Chitosan/polyethylene glycol diacrylate films as potential wound dressing material, *International Journal of Biological Macromolecules*, **43**, pp. 456–462.

74. Menda, J.P., Reddy, T., Deepika, R., Pandima Devi, M. and Sastry, T.P. (2014). Preparation and characterization of wound healing composites of chitosan, Aloe Vera and Calendula officinalis - A comparative study, *American Journal of Phytomedicine and Clinical Therapeutics*, **2**, pp. 061–076.

75. Nguyen, V.C., Nguyen, V.B. and Hsieh, M-F. (2013). Curcumin-loaded chitosan/gelatin composite sponge for wound healing application, *International Journal of Polymer Science*, Article ID 106570, pp. 1–7.

76. Kouchak, M., Ameri, A., Naseri, B. and Boldaji, S.K. (2014). Chitosan and polyvinyl alcohol composite films containing nitrofurazone: Preparation and evaluation, *Iranian Journal of Basic Medical Sciences*, **17**, pp. 14–20.

77. Zou, Q., Cai, B., Li, J., Jidong, Li. and Yubao, Li. (2017). *In vitro* and *in vivo* evaluation of the chitosan/Tur composite film for wound healing applications, *Journal of Biomaterials Sciences*, **28**, pp. 601–615.

78. Pandima Devi, M., Sekar, M., Chamundeswari, M., Moorthy, A., Krithiga, G., Selva Murugan, N. and Sastry, T.P. (2012). A novel wound dressing material – Fibrin–chitosan–sodium alginate composite sheet, *Bulletin of Materials Science*, **35**, pp. 1157–1163.

79. Dastan, T. and Turan, K. (2004). In vitro characterization and delivery of chitosan-DNA microparticles into mammalian cells, *Journal of Pharmacy and Pharmaceutical Sciences*, **7**, pp. 205–214.

80. Varshosaz, J., Sadrai, H. and Alinagari, R. (2004). Nasal delivery of insulin using chitosan microspheres, *Journal of Microencapsulation*, **21**, pp. 761–774.

81. Wang, W., Pei, X., Du, Y. and Li, Y. (2008). Quaternized chitosan/rectorite intercalative materials for a gene delivery system, *Nanotechnology*, **19**(37), pp. 375102.

82. Min, S-H., Park, K-C. and Yeom, Y.I. (2014). Chitosan-mediated non-viral gene delivery with improved serum stability and reduced cytotoxicity, *Biotechnology and Bioprocess Engineering*, **19**, pp. 1077–1082.

83. Nie, H., Lee, L.Y., Tong, H. and Wang, C.H. (2008). PLGA/chitosan composites from a combination of spray drying and supercritical fluid foaming techniques: New carriers for DNA delivery, *Journal of Controlled Release*, **129**, pp. 207–214.

84. Eroglu, E., Tiwari, P.M., Waffo, A.B., Miller, M.E., Vig, K., Dennis, V.A. and Singh, S.R. (2013). A nonviral pHEMA+chitosan nanosphere-mediated high-efficiency gene delivery system, *International Journal of Nanomedicine*, **8**, pp. 1403–1415.

85. Xu, B., Jiang, G., Yu, W., Liua, D., Liu, Y., Kong, X. and Yao, J. (2017). Preparation of poly(lactic-co-glycolic acid) and chitosan composite nanocarriers via electrostatic self assembly for oral delivery of insulin, *Materials Science and Engineering C*, **78**, pp. 420–428.

86. Ali, S., Sangi, L. and Kumar, N. (2017). Exploring antibacterial activity and hydrolytic stability of resin dental composite restorative materials containing chitosan, *Technology and Health Care*, **25**, pp. 11–18.

87. Samprasit, W., Kaomongkolgit, R., Sukma, M., Rojanarata, T., Ngawhirunpat, T., and Opanasopit, P. (2015). Mucoadhesive

electrospun chitosan-based nanofibre mats for dental caries prevention, *Carbohydrate Polymers*, **117**, pp. 933–940.

88. Ruan, Q., Siddiqah, N., Li, X., Nutt, S., and Moradian-Oldak, J. (2014). Amelogenin—chitosan matrix for human enamel regrowth: Effects of viscosity and supersaturation degree, *Connective Tissue Research*, **55**, pp. 150–154.

89. Perchyonok, V.T., Grobler, S., Zhang, S., Olivier, A. and Oberholzer, T. (2013). Insights into chitosan hydrogels on dentine bond strength and cytotoxicity, *Open Journal of Stomatology*, **3**, pp. 75–82.

90. Yang, X., Han, G., Pang, X. and Fan, M. (2011). Chitosan/collagen scaffold containing bone morphogenetic protein-7 DNA supports dental pulp stem cell differentiation *in vitro* and *in vivo*, *Journal of Biomedical Materials Research Part A*, **108**(12), 2519–2526.

91. Bonifacio, M.A., Cometa, S., Dicarlo, M., Baruzzi, F., de Candia, S., Gloria, A., Giangregorio, M.M., Mattioli-Belmonte, M. and De Giglio, E. (2017). Gallium-modified chitosan/poly(acrylic acid) bilayer coatings for improved titanium implant performances, *Carbohydrate Polymers*, **166**, pp. 348–357.

92. Hurt, A.P., Clavijo, A.A.R., Vine, G.J., Coleman, A.A. and Coleman, N.J. (2014). Antibacterial activity of a chitosan-PVA-Ag^+-Tobermorite composite for periodontal repair, *Conference Papers in Science*, **2014**, Article ID 684352, pp. 1–7.

93. Ghaee, A., Shariaty-Niassar, M., Barzin, J., Matsuura, T. and Ismail, A.F. (2016). Preparation of chitosan/cellulose acetate composite nanofiltration membrane for waste water treatment, *Desalination and Water Treatment*, **57**, pp. 14453–14460.

94. Elsabee M.Z. and Abdou E.S. (2013). Chitosan based edible films and coatings: A review, *Materials Science and Engineering C*, **33**, pp. 1819–1841.

95. Luo, Y. and Wang, Q. (2014). Recent development of chitosan-based polyelectrolyte complexes with natural polysaccharides for drug delivery, *International Journal of Biological Macromolecules*, **64**, pp. 353–367.

96. Al-Naamani, L., Dobretsov, S. and Dutta, J. (2016). Chitosan-zinc oxide nanoparticle composite coating for active food packaging applications, *Innovative Food Science & Emerging Technologies*, **38**, pp. 231–237.

97. Theapsak, S., Watthanaphanit, A. and Rujiravanit, A. (2012). Preparation of chitosan-coated polyethylene packaging films by DBD

plasma treatment, *ACS Applied Materials & Interfaces*, **4**, pp. 2474–2482.

98. Vasile, C., Darie, R.N., Sdrobiş, A., Pâslaru, E., Pricope, G., Baklavaridis, A., Munteanu, S.B. and Zuburtikudis, I. (2014). Effectiveness of chitosan as antimicrobial agent in LDPE/CS composite films as minced poultry meat packaging materials, *Cellulose Chemistry and Technology*, **48**, pp. 325–336.
99. Pereda, M., Dufresne, A., Aranguren, M.I. and Marcovich, N.E. (2014). Polyelectrolyte films based on chitosan/olive oil and reinforced with cellulose nanocrystals, *Carbohydrate Polymers*, **101**, pp. 1018–1026.
100. Buşilă, M., Muşat, V., Textor, T. and Mahltig, B. (2015). Synthesis and characterization of antimicrobial textile finishing based on Ag:ZnO nanoparticles/chitosan biocomposites. *RSC Advances*, 5, pp. 21562–21571.
101. Rajendran, R., Radhai, R. and Balakumar, C. (2012). Synthesis and characterization of neem chitosan nanocomposites for development of antimicrobial cotton textiles, *Journal of Engineered Fibers and Fabrics*, **7**, pp. 136–141.
102. Rajendran, K. and Sivalingam, T. (2013). Industrial method of cotton fabric finishing with chitosan–ZnO composite for anti-bacterial and thermal stability, *Industrial Crops and Products*, **47**, pp. 160–167.
103. Wang, X., Du, Y., Fan, L., Liu, H. and Hu, Y. (2005). Chitosan-metal complexes as antimicrobial agent: Synthesis, characterization and structure-activity study, *Polymer Bulletin*, **55**, pp. 105–113.
104. Anandhavelu, S. and Thambidurai, S. (2013). Single step synthesis of chitin/chitosan-based graphene oxide–ZnO hybrid composites for better electrical conductivity and optical properties, *Electrochimica Acta*, **90**, pp. 194–202.
105. Yang, G.J., Yang, X.Y. and Zhang, L. (2012). Counterionic biopolymers-reinforced bioactive glass scaffolds with improved mechanical properties in wet state, *Materials Letters*, **75**, pp. 80–83.
106. Lee, S.J., Heo, D.N., Moon, J.H., Ko, W.K., Lee, J.B., Bae, M.S., Park, S.W., Kim, J.E., Lee, D.H., Kim, E.C., Lee, C.H. and Kwon, I.K. (2014). Electrospun chitosan nanofibers with controlled levels of silver nanoparticles. Preparation, characterization and antibacterial activity, *Carbohydrates Polymers*, **111**, pp. 530–537.
107. Yuan, H., Chen, N., Lü, X. and Zheng, B. (2008). Experimental study of natural hydroxyapatite/chitosan composite on reconstructing bone defects, *Journal of Nanjing Medical University*, **22**, pp. 372–375.

108. Liuyun, J., Yubao, L. and Chengdong, X. (2009). Preparation and biological properties of a novel composite scaffold of nano-hydroxyapatite/chitosan/carboxymethyl cellulose for bone tissue engineering, *Journal of Biomedical Science*, **16**, pp. 1–10.

109. Matsuda, A., Ikoma, T., Kobayashi, H. and Tanaka, J. (2004). Preparation and mechanical property of core-shell type chitosan/calcium phosphate composite fiber, *Materials Science and Engineering C*, **24**, pp. 723–728.

110. Muzzarelli, R.A.A., Morganti, P., Morganti, G., Palombod, P., Palomboe, M., Biaginic, G., Belmontec, M.M., Giantomassic, F., Orlandif, F. and Muzzarellia, C. (2007). Chitin nanofibrils/chitosan glycolate composites as wound medicaments, *Carbohydrate Polymers*, **70**, pp. 274–284.

Chapter 5

Green Composites Based on Polyhydroxyalkanoates

Mufaro Moyo, Ajay Vasudeo Rane, Gbadeyan O. Joseph, and Krishnan Kanny
Composite Research Group, Department of Mechanical Engineering, Durban University of Technology, Durban–4000, South Africa
mufaromoyo@gmail.com

Green composites based on polyhydroxyalkanoates (PHAs) have received considerable attention over the years due to their major advantages of biodegradability and biocompatibility. This is in line with calls for environmentally friendly materials. Green PHA composites are suitable replacements for environmentally unfriendly nonbiodegradable composites. Processing of green composites based on polyhydroxyalkanoates can be done using either wet processing or melt processing. They can be processed into PHA-based microcomposites or PHA-based nanocomposites. Major applications of PHA-based green composites include biomedical materials, packaging materials, conventional commodity plastics, membrane technology, and in the electronics industry. These green

Green Micro- and Nanocomposites
Edited by Sabu Thomas, Abitha V. K., and Hanna J. Maria

ISBN 978-981-4968-79-9 (Hardcover), 978-1-003-42756-8 (eBook)
www.jennystanford.com

composites are not void of problems as they have drawbacks of being difficult to process and high cost of production. Current research focuses on addressing drawbacks associated with these green composites based on PHAs.

5.1 Introduction

Green composites based on polyhydroxyalkanoates are finding a lot of interest in academia, industry, and other fields. These composites have major unique properties of compostability and biodegradability, which are crucial in addressing the environmental and landfill challenges that are faced worldwide. These green composites have crucial applications in different key sectors such as biomedical, packaging, electronics, membrane technology, and chemical industries. This chapter covers processing, properties, applications, advantages, disadvantages, and recent developments of the green PHA composites.

5.2 Sources and Classes of Polyhydroxyalkanoate Composites

These are composites made by combining polyhydroxyalkanoates (PHAs) with reinforcing materials such as natural fibers or fillers. PHAs belong to the naturally occurring aliphatic biopolyester family that is biodegradable and biocompatible [1–5]. PHAs are produced from the resources shown in Fig. 5.1 using numerous Gram-positive and Gram-negative bacteria [2, 6].

PHAs are classified according to their carbon chain length [6–8]. Figure 5.2 shows three broad classes of PHAs that can be used in the production of polyhydroxyalkanoate composites.

scl-PHA, mcl-PHA and lcl-PHA contain up to C5, C6–C14, and above C14 monomers, respectively, and the properties of scl-PHA are close to conventional plastics and mcl-PHAs are regarded as elastomeric and rubbery [6, 8–10]. One common class of scl-PHAs is polyhydroxybutyrate (PHB) and the homopolymer of 3-hydroxybutyric acid has been substantially researched [6, 9]. Copolymers of PHA can be formed containing [6, 9, 11].

- 3-hydroxybutyrate (3HB),
- 3-hydroxyvalerate (3HV),
- 3-hydroxyhexanoate (3HH), and
- 4-hydroxybutyrate (4HB) monomers.

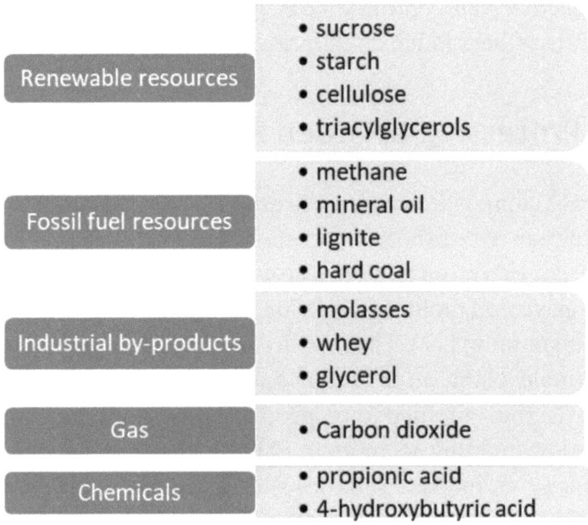

Figure 5.1 Resources used in the production of PHAs.

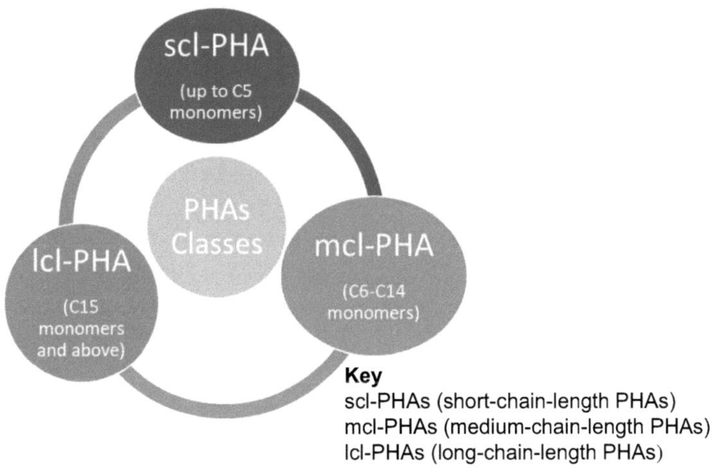

Figure 5.2 Classes of PHAs.

Only a few PHAs, predominantly PHB, poly(3-hydroxybutyrate), poly(3-hydroxybutyrate-*co*-3-hydroxyvalerate), and poly(3-hydroxybutyrate-*co*-3-hydroxyhexanoate), have proceeded to large-scale production and are used extensively to manufacture composites [1, 2, 6, 9, 11]. The composites are made by combining the PHAs with bioactive glass, natural fibers, glass-ceramic fillers or coatings, regenerated fibers (such as cellulose acetate), and hydroxyapatite.

5.3 Processing of Green PHA Composites

Green PHA composites can be processed using either melt processing technology or wet processing technology. Conventional processing methods for thermoplastics can be used in melt processing and these include extrusion molding, injection molding, compression molding, and melt spinning [17]. The mixture of fillers and polymers is firstly compounded using an extruder and pelletized. The compounded mixture is then molded into products using either compression or injection molding technique. Palletization can be done using a palletizer or an internal mixture such as Haake or Brabender [17]. The major challenges with melt processing are the low thermal degradation of PHA and fiber dispersion. There are two mechanisms for the thermal degradation of PHAs, namely, β-scission and E1cB, which lead to thermal degradation of PHAs at temperatures just above the melting point (170–200°C) [17]. It was discovered that at 190°C, the weighted-average molecular weight of (M_w) of PHB decreased from 490,000 to 72,000 for an exposure of 10 min and to 16,000 after another 20 min [17]. It was also discovered that when the filler material is basic or contains some alkali, degradation of PHAs would be a serious problem. Homogeneous dispersion of filler material in the matrix is of paramount importance in getting a final product with quality and consistent properties. Good composite properties are obtained if an extruder is used for compounding because it is effective in dispersing fillers into plastic materials.

In wet processing, a suitable solvent is used to dissolve PHAs. After dissolution, sodium chloride and filler material are added to the solution, which is then cast into a mold to form various shapes after evaporation of the solvent. The salt is leached by water to form a porous structure. Chloroform is the most common solvent for dissolving PHAs. Wet processing is often used for manufacturing 3D tissue engineering scaffolds [17]. One major drawback of wet processing is that there might be residual solvent in the final structure. Residual solvent might cause adverse side effects when the structure is used for biomedical applications [17]. Dry processing is the modified version of wet processing [17]. In dry processing, the mixture of salt and filler material is stirred strongly and then molded into different shapes using compression molding. However, it is difficult to achieve homogeneity of filler material distribution in the product. Infiltration processing is another modification of the wet processing method reported by Porter et al. [18]. Sintered HA scaffolds were soaked in a suspension of PHA particles and then the PHA particles were infiltrated into the scaffolds via centrifugation and the infiltrated PHA particles were melted to form a strong scaffold for tissue engineering applications [17, 18].

5.3.1 PHA-Based Microcomposites

PHA microcomposites are made by incorporating PHAs with micro fillers. The fillers, largely cellulosic materials such as fibers, are at the micro scale. Natural fiber micro fillers improve the tensile strength and stiffness of the resultant PHA microcomposites. There are a lot of research articles on PHA-based microcomposites that have been successfully manufactured for different applications. There is growing interest to mimic the characteristics of natural extracellular matrix (ECM) in the biomedical industry using different methods such as particle leaching, textile and phase separation technologies [14]. These methods resulted in the production of PHA microcomposites. These micro PHA microcomposite matrices have some advantages; however, since the diameters of the matrix fibers and pores are in the micron scale, they are far from the ECM [14]. Michael and his coworkers evaluated polyhydroxybutyrate-based natural fiber composites for tensile characterization [19].

5.3.2 PHA-Based Nanocomposites

PHA-based nanocomposites are made by incorporating PHAs with nanofillers. Distinguishable types of nanofillers based on dimensions in the nanometer scale are the following [20, 21]:

- One-dimensional, like layered silicates
- Two-dimensional, like nanotubes or whiskers
- Three-dimensional, like spherical silica particles

5.4 Properties of PHA Composites

PHA composites possess some interesting properties that dictate their application areas. Figure 5.3 shows the major properties of PHA composites.

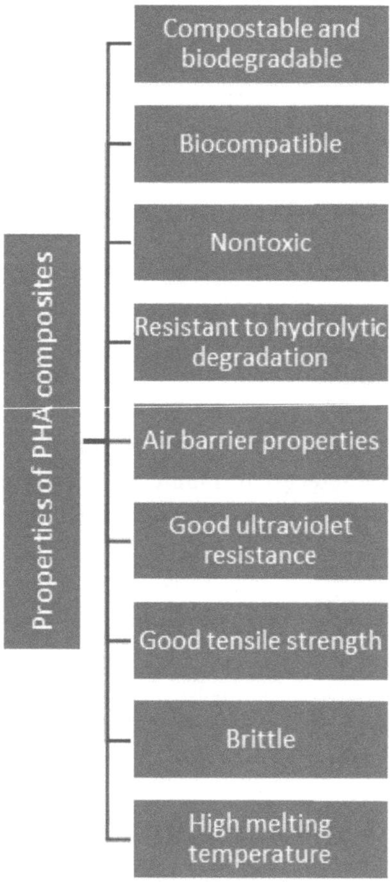

Figure 5.3 Major properties of PHA composites.

PHA composites have unique compostability and biodegradability properties that make these composites able to address materials post-use disposal and landfill challenges. They are compostable and biodegradable in different soil and aquatic environments. The composites are biocompatible making them suitable for applications in the medical field. PHAs are thermoplastic in nature hence their composites have good resistance to moisture and possess aroma or air barrier properties and are nontoxic hence they have applications in food packaging, medical industry, membrane technology, and conventional commodity plastics. They also have good ultraviolet resistance, which makes the composites resistant to UV light during use. PHA composites have good mechanical properties specifically high tensile strength and melting point, which are crucial for their performance during use. However, PHAs are intrinsically brittle, but the brittleness gets reduced during composite formation due to the addition of the fillers or reinforcing materials.

5.5 Advantages of PHA Composites

PHA composites have numerous advantages such as compostability, biodegradability, compatibility, and nontoxicity. However, they are not void of problems. This section discusses the major advantages and challenges associated with the PHA composites.

5.5.1 Advantages of Green PHA Composites

Green PHA composites can be explained as PHA biocomposites that are environmentally friendly. Biocomposites are made by incorporating fillers, mainly lignocellulosic fibers, into a biopolymer matrix. Figure 5.4 shows the advantages of green PHA composites.

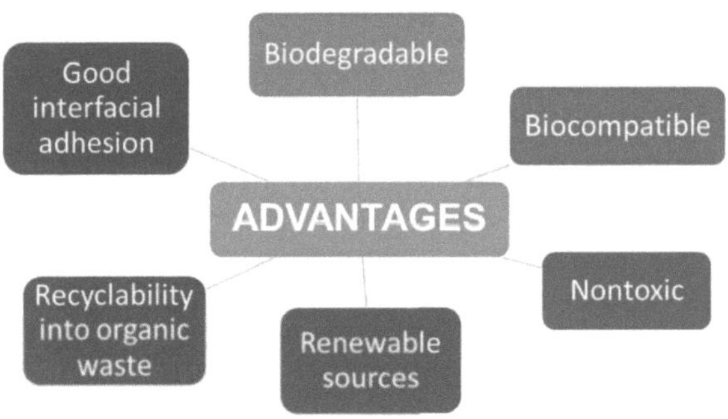

Figure 5.4 Advantages of green PHA composites.

Green PHA composites have a distinctive property of environmental degradability. Green PHA composites degrade in various environments under the soil, in compost, and in marine sediments [5, 6, 12–14]. Their biodegradation depends on different factors, notably those related to environmental conditions and those related to the green PHA composites themselves [6, 12, 14]. Figure 5.5 shows the factors that affect the biodegradation of green PHA composites.

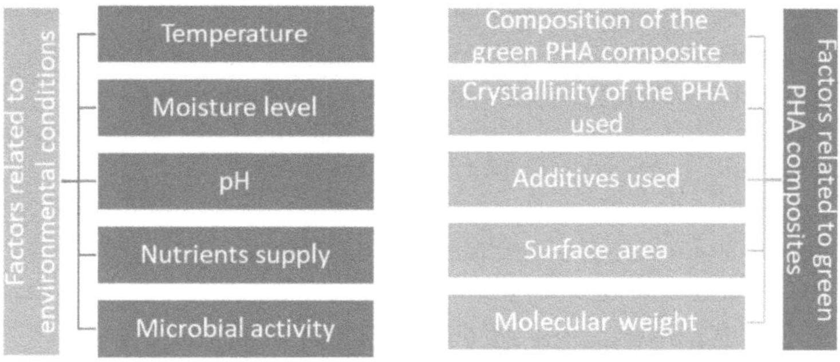

(a) Factors related to environmental conditions (b) Factors related to green PHA composites

Figure 5.5 Factors affecting the biodegradability of green PHA composites.

Degradation at the surface was observed on electron microscopy due to enzymatic hydrolysis (morphological erosion), but the molecular weight of the degraded PHAs remained unaltered [12]. PHAs in green PHA composites are effectively degraded by PHB depolymerase enzymes into water-soluble monomers and oligomers. [2]. These enzymes that hydrolyze the esters are secreted by microorganisms. In addition, the microorganisms discharge extracellular degrading enzymes that disintegrate solid PHA by hydrolysis into hydroxy acids, which are a source of carbon for growth [2, 6, 9, 12]. PHAs in green PHA composites can be degraded at a fast rate of between 3 and 9 months by microorganisms, which use their own secreted PHA depolymerase [9]. Polymer composition and crystallinity are also important in the biodegradation of green PHA composites. PHB monomers containing copolymers were observed to degrade faster than PHB and 3HB-*co*-3HV copolymers [6]. A broad range of temperatures can be used in composting green PHA composite materials. Research showed that temperatures of around

60°C with moisture levels at 55% could be used in composting PHA and it was observed in seven weeks that about 85% of PHA had degraded [6]. In aquatic environments with temperatures not exceeding 6°C, PHA was reported to degrade within 64 weeks [6]. PHA composites biodegradation results in carbon dioxide and water for aerobic conditions and carbon dioxide and methane for anaerobic conditions [6].

Biocompatibility is another unique property of green PHA composites. Green PHA composites are compatible with soft body tissues or blood. This is critical in medical applications. In vitro and in vivo analyses showed compatibility between green PHA composites and cartilage and bone tissue, cell lines, and blood [1, 5, 14, 15]. There are several reports on green PHA composites fabricated using PHAs and nonwoven fabrics, film materials, and sutures for the application in the medical field [1, 5, 14]. Green PHA composites are nontoxic. This is a very important property as far as the application of green PHA composites is concerned. PHAs are ubiquitous, interactive, solvating biopolymers that are involved in important physiological functions. Research showed that derivatives of 3-hydroxy acids could be used as a source of ketone for increasing ketone in the blood of mammals because they are nontoxic [11]. Green PHA composite materials used in the medical field are not toxic, and they provide therapeutical and nutritional benefits instead.

Other advantages of green PHA composites are that they can be recycled into organic waste. Green PHA composites can be completely recycled into organic wastes [2]. Green PHA composites also have another advantage of being produced from renewable sources. The PHAs and the reinforcing materials or fillers used in producing green PHA composites are obtained from renewable resources unlike petroleum-based composites that are produced from nonrenewable resources, which are getting depleted. PHAs are obtained from renewable carbon sources, which makes them materials of choice in green composite production [5]. Last but not least, green PHA composites are made with materials that have good interfacial adhesion. The polar nature of PHAs, specifically PHB, enables them to have better adhesion to natural fibers than polyolefins [13]. This means that the resulting green PHA composites have better mechanical properties than those obtained with olefins.

5.5.2 Disadvantages of Green PHA Composites

Green PHA composites are not void of problems. Most of the challenges are due to the problems associated with the PHAs. Some of the drawbacks associated with green PHA composites are shown in Fig. 5.6.

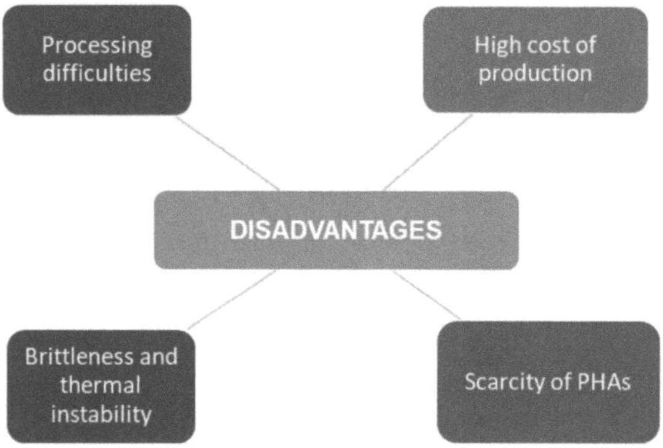

Figure 5.6 Disadvantages of green PHA composites.

- One of the concerns with green PHA composites is the difficulties associated with the production of PHAs. PHAs like PHB, which have a high melting temperature of approximately 170°C, are hard to process as these temperatures are very near to their degradation temperature [16]. Rigorous and careful controls are needed during processing to maintain the original molecular weight and properties [2, 13].
- Green PHA composites are expensive to produce due to the expensive ways associated with the production of PHAs. The overall economics of the cost of producing green PHA composites is affected by the intracellular recovery process of PHA [16]. Intensive research is being done with cheap carbon sources and genetic engineering so as to optimize the production and applications of PHAs [2, 13].
- There is a scarcity of PHA-producing microorganisms for large-scale production of PHA. Only a few microorganisms among the numerous microorganisms can synthesize PHAs on large scale. A few include the following [2]:

- ✓ *Cupriavidus necator* (formerly known as *Ralstonia eutropha* or *Alcaligenes eutrophus*)
- ✓ *Alcaligenes latus*
- ✓ *Azotobacter vinelandii*
- ✓ *Pseudomonas oleovorans*
- ✓ *Paracoccus denitrificans*
- ✓ *Protomonas extorquens*
- ✓ Recombinant *E. coli*

This is a major hindrance to the production of high quantities of green PHA composites.

- Brittleness and thermal instability of PHB in melt processing temperatures above its melting point give a narrow processing window [13, 16]. This instability makes processing difficult at temperatures above melting points. The thermochemical properties of PHAs have been observed to be related to their structure with brittle character and low flexibility. PHB and PHBV materials, for instance, have been shown to fracture at less than 10% elongation [16].

5.6 Applications of PHA Composites

PHA composites have a wide variety of applications in biomedical, chemical, electronics, plastics, and packaging industries. They are also used in membrane technology and as chiral precursors. The following sections discuss the various applications of the PHA composites.

5.6.1 Applications of PHA Microcomposites

PHA microcomposites have applications in areas such as the biomedical field, packaging, conventional commodity plastics and as chiral precursors. Figure 5.7 shows the applications of PHA microcomposites.

In the biomedical field, PHA microcomposites are used in the production of scaffolding materials, surgical materials, drug release, sutures, wound dressing, cardiovascular grafts, stents, patches, drug release, and guided tissue/regenerative devices. This is because the PHA microcomposites are biocompatible and nontoxic. In vivo studies showed that PHB and hydroxyapatite blends can be used for bone scaffolding when treating bone defects, pulmonary valve leaflets and arteries can be produced from copolymers of polyglycolic acid and PHB, PHA-based composites can be used as heart valve scaffold, graft matrix for neuronal generation after injuries to the spinal cord, and in rats, PHB films were found to provide scaffolding for large bowel defects patching [1, 6, 11, 15, 22–25]. PHA microcomposites are also used as biodegradable transporters of lengthy sustained dosages of drugs, medicinal, hormonal, pesticides and herbicides and as osteosynthetic materials in the stimulation of bone growth [6, 10, 15]. PHB bioactive and biodegradable composites are used in the replacement and revivification of hard tissue and engineering scaffolds for bone tissues [11]. In the packaging field, PHA microcomposites are used in the production of bags, containers, and paper coatings for food and beverage packaging. They are favorable for these applications because they are biodegradable and have low permeability to water and gases such as carbon dioxide and oxygen without residues of catalysts and water solubility. PHA microcomposites are also used for making conventional commodity plastics because of their non-toxicity and environmentally friendliness. Last, but not least, PHA microcomposites are also used as chiral precursors for making optically active compounds through chemical synthesis and synthesis of antibiotics.

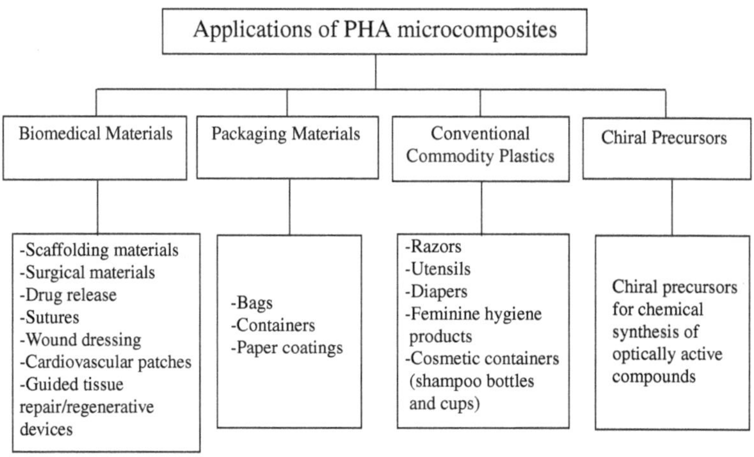

Figure 5.7 Applications of PHA microcomposites.

5.6.2 Applications of PHA Nanocomposites

PHA nanocomposites have applications in areas such as the medical industry, food packaging, membrane technology, and electronics industry. More applications for PHA nanocomposites are being explored through relentless research being done. Figure 5.8 shows some of the applications of PHA nanocomposites.

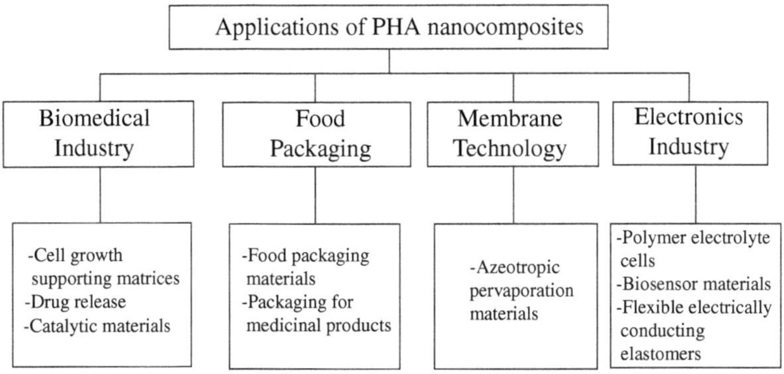

Figure 5.8 Applications of PHA nanocomposites.

PHA nanocomposites are also used in the biomedical industry to make materials for supporting cell growth [14]. Li et al. successfully fabricated PHA nanofiber matrices that supported cell attachment and viability [14]. Electrospun PHBV nanofabric was found to provide structures for tissue engineering cell culture surfaces [11]. PHA nanocomposites such as pseudo-PHA granules (nanoparticles) fabricated in vitro are also used as effective drug delivery devices [15]. Zhang et al. reported the use of PHA/folate ligands to carry out anticancer therapy [21]. Metal intercalated nanoparticles exhibit catalytic properties and hence are used to make catalytic materials. One example of these PHA nanocomposite catalytic materials is PHB with silver sulfide nanoparticles. PHA nanocomposites are also used in packaging for food materials and packaging for medicinal products. PHA nanocomposites containing silver nanoparticles exhibit antimicrobial activities that inhibit the growth of post-processing microbial contaminants. This extends the shelf life of food and improves food safety.

Metal intercalated nanoparticles exhibit electrical and sensing properties. Consequently, PHA nanocomposites containing these metal intercalated nanoparticles are used in the electronics industry. PHA nanocomposites are used for making polymer electrolyte cells, biosensor materials, and flexible electrically conducting elastomers. These are shape memory PHA nanocomposite materials. Poly(3-hydroxyoctanoate-*co*-3-hydroxyundecenoate) composited with silsesquioxane nanoparticles, for instance, are used for biosensors [16]. Solid polymer electrolyte cells made using PHA nanocomposites are being produced as an alternative to liquid electrolyte, which has drawbacks of being flammable, gas formation due to overcharging, and release of toxic and hazardous materials into the environment following end-of-life disposal [16]. PHA carbon nanotube nanocomposites are used for making flexible electrically conducting elastomers for neural applications [16]. PHA nanocomposites are also used in membrane technology. They are used for making membrane materials for the separation of substances. PHB with multi-walled carbon nanotubes/chitosan matrix as nanoparticles, for instance, is used in azeotropic pervaporation [16]. PHA nanocomposite membranes are also used in biomedical applications.

5.7 Recent Developments in Green PHA Composites

Advances in metabolic and genetic engineering have opened opportunities to increase the yields of PHA, produce tailor-made PHAs, and use new bacterial substrates. Reddy et al. reported the advances that brought opportunities to introduce new metabolic pathways that broadened the utilizable substrate range, making it possible to produce tailor-made PHA and increased yields [2, 6]. Research also focuses on the cost-effective, efficient, and high production of PHA [2, 26]. Suitable sources of carbon are selected for efficient and low-cost production of PHA and subsequent PHA products. Some of the carbon sources on which research is being conducted are sugars, starches, and alcohols. There are a number of reported research successes in producing PHAs from sugar, starches, and alcohols [2]. Also in this domain, some of the recent developments are on using some industrial by-products as carbon sources for producing PHAs. Some of the researched industrial by-products include molasses, fatty acids, and vegetable oils [2]. There are studies on understanding host-plasmid interactions for the development of stable plasmids for recombinant bacteria to surpass the wild-type bacteria [6]. There is also the development of transgenic plants harboring the microbial PHA biosynthesized genes [6]. Production of PHA in transgenic plants results in high volume production of the PHA, which increases the supply toward the currently very high demand for these polymers.

Research to control the fermentation conditions and application of mathematical models is being conducted so as to improve the production of PHB and use suboptimal procedures by maintaining constant nutrient concentration [16]. Online feeding for fed-batch cultivation is being developed to augment the efficient production of PHAs. A multivariable control strategy for alcohol concentration during the production process of PHBV has been developed [2].

5.8 Conclusion and Perspectives

Green composites based on polyhydroxyalkanoates are receiving great attention and acceptability as they have great potential of replacing nonbiodegradable composites in some applications. They are promising materials that would address issues of environmental sustainability and waste disposal challenges as some of the waste can act as precursors in the production of PHAs. Their major advantages of biodegradability and biocompatibility make them kind of unique composite materials with applications in the biomedical field, packaging film, conventional plastics, membrane technology, and the electronics field. Micro- and nanocomposites of PHA can be fabricated using the wet or melt processing technique. However, their cost of production is currently high and their processing has some technical difficulties that need to be addressed. Continued research on addressing the challenges associated with green composites based on PHAs would reduce the cost of production and improve production volumes. Considerable research on the potential application in biomedical applications was done largely on animals. It is of paramount importance to mention that this should now be extended to human beings for the benefit of humanity.

Processing techniques dedicated to green PHA composites should be developed so as to deal with the processing difficulties currently being encountered.

Abbreviations

PHA	Polyhydroxyalkanoate
scl-PHAs	Short-chain-length polyhydroxyalkanoates
mcl-PHAs	Medium-chain-length polyhydroxyalkanoates
lcl-PHAs	Long-chain-length polyhydroxyalkanoates
PHB	Polyhydroxybutyrate
3-HB	3-Hydroxybutyrate
3-HV	3-Hydroxyvalerate
3-HH	3-Hydroxyhexanoate
4HB	4-Hydroxybutyrate
PHBV	Poly(hydroxybutyrate-*co*-hydroxyvalerate)
M_w	Molecular weight
ECM	Extracellular matrix
CNTs	Carbon nanotubes
CNFs	Carbon nanofibers
SGBG	Sol-gel bioactive glass
P(HB-*co*-HV)	Poly(hydroxybutyrate-*co*-hydroxyvalerate)

References

1. S. K. Misra, S. P. Valappil, I. Roy, and A. R. Boccaccini, "Polyhydroxyalkanoate (PHA)/inorganic phase composites for tissue engineering applications," *Biomacromolecules*, vol. 7, pp. 2249–2258, 2006.

2. S. Chanprateep, "Current trends in biodegradable polyhydroxyalkanoates," *Journal of Bioscience and Bioengineering*, vol. 110, pp. 621–632, 2010.

3. H.-K. Ma, M.-M. Liu, S.-Y. Li, Q. Wu, J.-C. Chen, and G.-Q. Chen, "Application of polyhydroxyalkanoate (PHA) synthesis regulatory protein PhaR as a bio-surfactant and bactericidal agent," *Journal of Biotechnology*, vol. 166, pp. 34–41, 2013.

4. A. Saadat, S. Karbasi, A. A. B. Ghader, and M. Khodaei, "Characterization of biodegradable P3HB/HA nanocomposite scaffold for bone tissue engineering," *Procedia Materials Science*, vol. 11, pp. 217–223, 2015.

5. E. E. Tănase, M. E. Popa, M. Râpă, and O. Popa, "PHB/cellulose fibers based materials: physical, mechanical and barrier properties," *Agriculture and Agricultural Science Procedia*, vol. 6, pp. 608–615, 2015.
6. C. S. K. Reddy, R. Ghai, Rashmi, and V. C. Kalia, "Polyhydroxyalkanoates: an overview," *Bioresource Technology*, vol. 87, pp. 137–146, 2003.
7. A.-L. Chung, G.-D. Zeng, H.-L. Jin, Q. Wu, J.-C. Chen, and G.-Q. Chen, "Production of medium-chain-length 3-hydroxyalkanoic acids by β-oxidation and phaC operon deleted Pseudomonas entomophila harbouring thioesterase gene," *Metabolic Engineering*, vol. 17, pp. 23–29, 2013.
8. S. Khanna and A. K. Srivastava, "Recent advances in microbial polyhydroxyalkanoates," *Process Biochemistry*, vol. 40, pp. 607–619, 2005.
9. P. Suriyamongkol, R. Weselake, S. Narine, M. Moloney, and S. Shah, "Biotechnological approaches for the production of polyhydroxyalkanoates in microorganisms and plants – a review," *Biotechnology Advances*, vol. 25, pp. 148–175, 2007.
10. S. J. Park, J.-I. Choi, and S. P. Lee, "Short-chain-length polyhydroxyalkanoates: synthesis in metabolically engineered Escherichia coli and medical applications," *Journal of Microbiology and Biotechnology*, vol. 15, pp. 206–215, 2005.
11. G.-Q. Chen and Q. Wu, "The application of polyhydroxyalkanoates as tissue engineering materials," *Biomaterials*, vol. 26, pp. 6565–6578, 2005.
12. K. Sudesh, H. Abe, and Y. Doi, "Synthesis, structure and properties of polyhydroxyalkanoates: biological polyesters," *Progress in Polymer Science*, vol. 25, pp. 1503–1555, 2000.
13. N. M. Barkoula, S. K. Garkhail, and T. Peijs, "Biodegradable composites based on flax/polyhydroxybutyrate and its copolymer with hydroxyvalerate," *Industrial Crops and Products*, vol. 31, pp. 34–42, 2010.
14. X.-T. Li, Y. Zhang, and G.-Q. Chen, "Nanofibrous polyhydroxyalkanoate matrices as cell growth supporting materials," *Biomaterials*, vol. 29, pp. 3720–3728, 2008.
15. C. J. Brigham and A. J. Sinskey, "Applications of polyhydroxyalkanoates in the medical industry," *International Journal of Biotechnology for Wellness Industries*, vol. 1, pp. 53–60, 2012.

16. I. Roy and Vishakh P. M. (eds.), *Polyhydroxyalkanoate (PHA) Based Blends, Composites and Nanocomposites*. Cambridge, UK: The Royal Society of Chemistry, 2015.
17. Inamuddin (ed.), *Green Polymer Composites Technology: Properties and Applications*. CRC Press, 2016.
18. M. M. Porter, S. Lee, N. Tanadchangsaeng, and M. J. Jaremko, Porous hydroxyapatite-polyhydroxybutyrate composites fabricated by a novel method via centrifugation, pp. 63–71, in *Mechanics of Biological Systems and Materials* (B. C. Prorok et al., Eds.), The Society for Experimental Mechanics, 2013.
19. M. A. Gunning, L. M. Geever, J. A. Killion, et al., "Mechanical and biodegradation performance of short natural fibre polyhydroxybutyrate composites," *Polymer Testing*, vol. 32, pp. 1603–1611, 2013.
20. M. A. Diez-Pascual, "Packaging nanocomposites based on ZnO-reinforced polyhydroxyalkanoates" in *Food Packaging: Nanotechnology in the Agri-Food Industry*, Volume 7. A. Grumezescu, Mihai, Ed. Academic Press, 2017.
21. Q. Zhang, Q. Liu, J. E. Mark, and I. Noda, "A novel biodegradable nanocomposite based on poly (3-hydroxybutyrate-co-3-hydroxyhexanoate) and silylated kaolinite/silica core-shell nanoparticles," *Applied Clay Science*, vol. 46, pp. 51–56, 2009.
22. Q. Wu, Y. Wang, and G.-Q. Chen, "Medical application of microbial biopolyesters polyhydroxyalkanoates," *Artificial Cells, Blood Substitutes, and Biotechnology*, vol. 37, pp. 1–12, 2009.
23. W. J. E. M. Habraken, J. G. C. Wolke, and J. A. Jansen, "Ceramic composites as matrices and scaffolds for drug delivery in tissue engineering," *Advanced Drug Delivery Reviews*, vol. 59, pp. 234–248, 2007.
24. S. F. Williams, D. P. Martin, D. M. Horowitz, and O. P. Peoples, "PHA applications: addressing the price performance issue: I. Tissue engineering," *International Journal of Biological Macromolecules*, vol. 25, pp. 111–121, 1999.
25. K. Zhao, Y. Deng, J. Chun Chen, and G.-Q. Chen, "Polyhydroxyalkanoate (PHA) scaffolds with good mechanical properties and biocompatibility," *Biomaterials*, vol. 24, pp. 1041–1045, 2003.
26. T. Keshavarz and I. Roy, "Polyhydroxyalkanoates: bioplastics with a green agenda," *Current Opinion in Microbiology*, vol. 13, pp. 321–326, 2010.

Chapter 6

Green Composites Based on Poly(Lactic Acid)

Ajay Vasudeo Rane,[a,b] Deepti Yadav,[c] Anupam Glorious Lobo,[d] Sabana Ara Begum,[b,e] P. Santhana Gopala Krishnan,[e] Aswathy J. S.,[f] and Krishnan Kanny[b]

[a]*Materials Research Laboratory, Department of Mechanical Engineering, KMEA Engineering College, Cochin, Kerala, India*

[b]*Composite Research Group, Department of Mechanical Engineering, Faculty of Engineering and Built Environment, Durban University of Technology, Durban-4000, South Africa*

[c]*Department of Biotechnology and Food Science, Durban University of Technology, Durban-4000, South Africa*

[d]*School of Chemical Sciences, Mahatma Gandhi University, Kottayam-686 560, Kerala, India*

[e]*Advance Polymer Design and Development Research Laboratory (APDDRL), School for Advance Research in Polymers (SARP), Central Institute of Petrochemicals Engineering and Technology (CIPET), Bengaluru, India*

[f]*School of Nanoscience and Nanotechnology, Mahatma Gandhi University, Kottayam-686 560, Kerala, India*

avr.me@kmeacollege.ac.in

6.1 Introduction

Biodegradable polymers (b-p) are divided into petroleum-based biodegradable polymers (P/b-p) and microorganism-

Green Micro- and Nanocomposites
Edited by Sabu Thomas, Abitha V. K., and Hanna J. Maria

ISBN 978-981-4968-79-9 (Hardcover), 978-1-003-42756-8 (eBook)
www.jennystanford.com

based biodegradable polymers (M/b-p) [1]. Production of P/b-p involves the emission of greenhouse gases, and the bioactivity of microorganisms is utilized to synthesize monomers from plant products, which are further used to synthesize polymers [1]. PLA is an aliphatic polyester made up of lactic acid building blocks [2]. Poly(lactic acid) (PLA) is produced by utilizing the activity of microorganisms. PLA is a biodegradable and compostable thermoplastic derived from renewable plant sources, and its use was limited to medical applications owing to bioabsorbable characteristics. Successful attempts to synthesize high molecular weight PLA economically have expanded the use of PLA for consumer and packaging applications [3]. As this chapter is focused on composites based on PLA, the synthesis of PLA is not discussed in detail – in short PLA is obtained commercially by ring-opening polymerization of lactide and condensation polymerization of lactic acid as shown in Fig. 6.1. Green movement stimulated the use of biobased polymers [1]. Hence, the word "Green composites" in the title of the chapter emphasizes the composites fabricated by making use of biobased polymers. As mentioned above, PLA is a polymer made from renewable sources and has the potential to reduce our dependency on conventional plastics made from fossil-based resources. Inventions and publications in studies related to PLA have developed tremendously, as evident from the graph shown in Fig. 6.2.

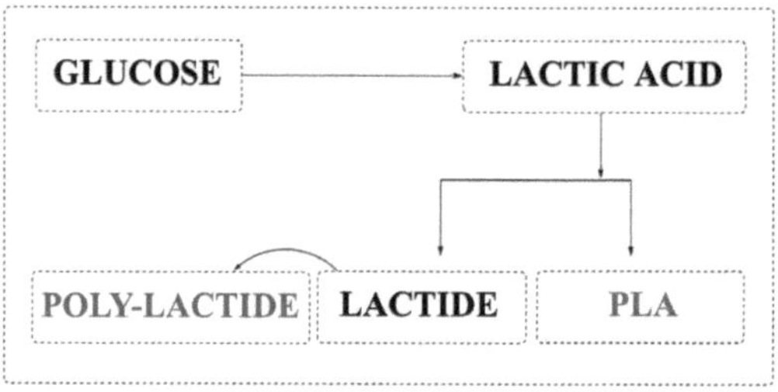

Figure 6.1 General routes for poly(lactic acid) synthesis and production.

6.1.1 Advantages and Disadvantages of PLA

Biobased nature, biodegradability, biocompatibility, excellent transparency, ease of processing, high strength, and stiffness are advantages of PLA. The high production cost and inherent brittleness of PLA are the disadvantages [4–9].

6.1.2 PLA Composites

Growing attention to PLA is interrelated to some distinctiveness that is deficient in other polymers, specifically concerning renewability, biocompatibility, processability, and energy saving. PLA is derivative from renewable and biodegradable resources, and its degradation products are non-pollutant and nontoxic. Therefore, PLA is a substitute for petrochemical plastics. Furthermore, PLA has several bio-applications, such as a biodegradable matrix for surgical implants and in drug delivery systems [10]. On the other hand, for structural use, it is required that some of its properties be improved, namely, its toughness. To improve toughness, additive approaches such as blending with other polymers and adding fillers are practiced, but these approaches do not give a balance of stiffness and toughness. Hence, a designed processing technique with a proper selection of additives would help in getting a balance of stiffness and toughness in PLA composites. In this chapter, PLA composites filled with particulate fillers are discussed. Let us now look into the recent literature on various fillers added in PLA.

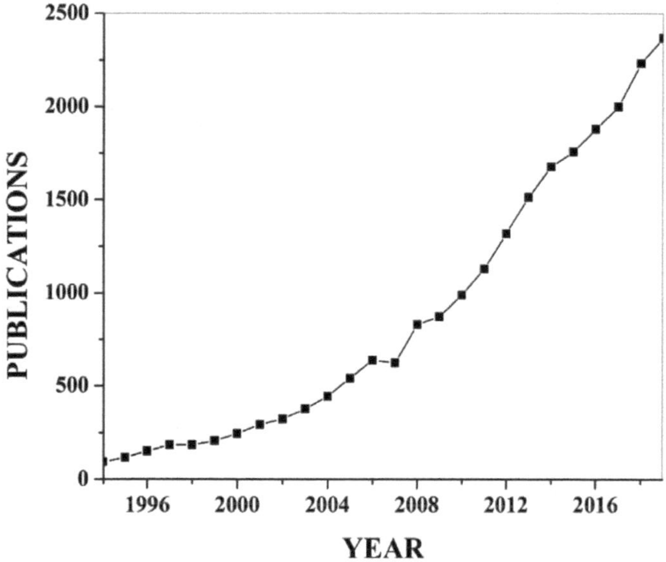

Figure 6.2 Research publications on PLA (1994–2019). Data Source: Web of Science (South Africa).

The rotational jet spinning technique was used by Hongthipwaree et al. to load neomycin into biodegradable fibers of PLA blended with polycaprolactone (PCL) to study drug release and drug loading. The effect of the rotational speed of the jet and the effect of blend ratio on the percentage of porosity and morphology were studied. It was confirmed that increasing the rotational speed of the jet decreased the diameter of the fiber, whereas the increasing percentage of PLA in the blend increased the fiber diameter and decreased the percentage of porosity. In vitro studies done by Hongthipwaree et al. concluded that a higher specific area increases the efficiency of drug release and drug loading of fibers [4].

Ghorbani et al. and his group from Iran loaded tetracycline hydrochloride on Zein/gum tragacanth/PLA nanofibers (composites) scaffolds to study wound healing and anti-inflammatory effect; in this work, Ghorbani et al. studied the effect of varied blending ratio of Zein and gum tragacanth on the diameter of Zein/gum tragacanth/PLA nanofibers. Increasing Zein content increased the diameter of nanofibers. Successful loading of tetracycline hydrochloride on Zein/gum tragacanth/PLA nanofibers was confirmed by Fourier transform infrared spectroscopy (FTIR) and thermal gravimetric analysis (TGA). The addition of tetracycline hydrochloride in Zein/gum tragacanth/PLA composites increased the tensile strength, degree of swelling, and porosity. Ghorbani et al. confirmed good antibacterial properties and their suitability for skin tissue engineering – with proliferation and adhesion against NIH-3T3 fibroblast cells [5].

Zhao et al. from China developed a new precursor to enhance the compatibility between PLA and poly(butylene adipate-*co*-terephthalate) (PBAT). The new precursor synthesized with two epoxy groups was 1,3-bis(2,3-epoxypropyl)-*s*-triazine-2,4,6-trione (BETT). The increasing compatibility between PLA and PBAT was due to the chain extensions of BETT, which hindered the crystallization of PLA. Scanning electron microscopic (SEM) images changed to co-continuous structure from sea-island structure, confirming the interfacial compatibility between PLA and PBAT with increasing concentration of BETT. Improvement in mechanical properties also confirmed the increase in interfacial compatibility between PLA and PBAT on the addition of BETT. Further PLA/PBAT/BETT composites demonstrated excellent antibacterial properties [6].

Zeng et al. studied thermal, mechanical, and electromagnetic interference (EMI) shielding properties of graphene nanoplatelets reinforced PLA/poly(ethylene oxide) (PEO) nanocomposites. Percolating 3D network was confirmed by rheological analysis and SEM images for PLA/PEO/graphene nanoplatelets nanocomposites with loading above 6 wt%. An increase in tensile modulus and tensile strength with acceptable tensile elongation was observed for PLA/PEO/graphene nanoplatelets nanocomposites with the loading of 15 wt%. An increase in initial degradation temperature and EMI shielding properties were observed for PLA/PEO/graphene nanoplatelets nanocomposites with increasing concentrations of graphene nanoplatelets. The addition of 1 wt% of carbon nanotubes (CNTs) remarkably improved the thermal, mechanical, and EMI shielding properties of PLA/PEO/graphene nanoplatelets nanocomposites [7].

Moradi and Yeganeh from Iran prepared highly toughened polymer nanocomposites of PLA/ethylene-*co*-vinyl acetate (EVA)/hydrophobic nanosilica and maleic anhydride grafted poly(styrene-*b*-ethylene-*co*-butylene-*b*-styrene) (SEBS-*g*-MA) block copolymer using melt blending. Mechanical properties, morphology, and linear rheology on fabricated PLA/ethylene-*co*-vinyl acetate (EVA)/hydrophobic nanosilica and SEBS-*g*-MA block copolymer composites were studied concerning the effect of nanosilica and SEBS-*g*-MA block copolymer. Transmission electron microscopic (TEM) images along with thermodynamic and kinetic analysis confirmed selective localization of nanosilica in the PLA and some at the interface of PLA-EVA. A decrease in average droplet size confirmed the increasing adhesion between the PLA-EVA on the addition of nanosilica; further addition of SEBS-*g*-MA block copolymer changed the spherical droplets to a cylindrical shape. Moradi and Yeganeh performed differential scanning calorimetry (DSC) analysis and confirmed a reduction in the crystallization rate of PLA on the addition of EVA; the addition of nanosilica increased the crystallization rate of PLA in PLA-EVA-nanosilica composites, whereas the addition of SEBS-*g*-MA block copolymer hindered the crystallization rate of PLA in PLA/EVA/hydrophobic nanosilica and SEBS-*g*-MA block copolymer nanocomposites. Simultaneous addition of nanosilica and SEBS-*g*-MA block copolymer exhibited excellent toughness property of PLA/EVA/hydrophobic nanosilica and SEBS-*g*-MA block copolymer nanocomposites [8].

Tessanan et al. were successful in their attempt to improve the impact performance of PLA by adding modified natural rubber (NR). NR was modified using a two-step process – hydrogenation followed by epoxidation. Modified NR was then melt blended with PLA, thereby increasing the impact performance and elongation at break. Improvement in tensile strength and tensile modulus of the PLA-modified NR blend was observed in comparison to PLA and PLA-unmodified NR. Morphological evidence confirmed the presence of fine rubber particles and fibrils on the impact-fractured specimens of the PLA-modified NR blend, which confirms interfacial adhesion between PLA-modified NR [9].

Tang et al. fabricated interfacial metalized segregated composites and conventional segregated composites using PLA/poly(ε-caprolactone) (PCL)/multiwalled carbon nanotubes (MWCNT) to study enhancement in EMI shielding property. Tang et al. confirmed segregated structure to be one of the reasons for the improvement in EMI shielding, wherein the conventional segregated structure showed a decrease in EMI shielding. Improvement in EMI shielding can be correlated to segregated structure and highly conductive interface in interfacial metalized segregated composites [11].

Lithium-ion battery separator membranes using PLA were synthesized by Barbosa et al. with a motive to reduce the impact on the environment by improving their recyclability. PLA in 8, 10, and 12 wt% in dichloromethane and dimethylformamide was studied for morphology, thermal, electrical, and electrochemical properties. It was confirmed that the morphology and porosity of the lithium-ion battery separator membranes using PLA could be tuned by varying PLA concentration in a solvent. 10 wt% PLA in dichloromethane and dimethylformamide showed the best ionic conductivity and a high discharge capacity value [12].

Rarima et al. from India prepared PLA/gelatin foams by non-solvent induced phase separation for biomedical applications, wherein hydrophilic, biocompatible, and non-foamable properties of gelatin were applied to generate porous patterns in a mechanically stable PLA matrix. Controlled release of metformin hydrochloride was observed due to the presence of gelatin and porosity. The presence of gelatin influences the crystallinity rate of PLA, affecting the rate of degradation of PLA foams. A high degree of biodegradability and cell viability of PLA-gelatin foams makes them suitable for biomedical materials [13]. In another work by Rarima et al., porous PLA/nano silver composite membranes were fabricated for catalytic reduction of 4-nitrophenol. A simple technique wherein the capping of silver nanoparticles by PLA is executed through the reduction of silver ions by glucose is reported. This method highlights the possible in situ reduction and the stabilizing effect through reaction occurring at the water-dichloromethane interface to develop a composite for the reduction of 4-nitrophenol. This method allowed to have control over the particle size of the nano silver and the coating thickness of PLA over nano silver. Nanocomposites loaded with 20% of silver nitrate were able to show enhanced catalytic activity toward the reduction of sodium borohydride reduction of 4-nitrophenol up to four cycles of operation. Complete reduction of 4-nitrophenol in 8 min was observed for new nano silver/PLA composites, with the possibility for industrial applications [14].

Chronaki et al. fabricated PLA microcapsules for drug delivery applications using solid-state polymerization. Encapsulation of active compounds, that is, drug in polymer carriers, protects and improves the physical characteristics of the active compound. The molecular weight and crystallinity of polymers determine the barrier properties of the polymeric carrier and the efficiency of the polymeric carriers [15].

Yan et al. from China fabricated biocompatible and biodegradable PLA/nano-hydroxyapatite (n-HAP) composites for bone repair application. PLA/n-HAP microspheres were synthesized using the emulsion solvent evaporation technique and then the PLA/n-HAP composites were fabricated using the laser powder bed fusion technique. Effects of various parameters of the emulsion solvent evaporation technique on microspheres size and morphology were studied. The effect of parameters involved during laser powder bed fusion technique on physicochemical and biological properties of fabricated parts was also studied. Composites filled with 10 wt% of n-HAP facilitated the osteogenic differentiation of rat stem cells [16].

Dubinenko et al. fabricated PLA/hydroxyapatite (HAP) composites by fused deposition modeling 3D printing for biomedical applications. 12.5, 25, and 50 wt% of HAP were selected for fabrication of scaffold. Dubinenko et al. used a series of steps in ball milling to disperse HAP within the PLA matrix; finally, extrusion was used to fabricate the PLA/HAP composites and tested for morphological and mechanical properties. Young's modulus increases with an increase in the concentration of HAP. The morphological analysis confirmed that an increase in HAP content increases the surface roughness, thereby increasing wettability and adhesion of cells on the fabricated PLA/HAP scaffolds [17].

Wang et al. grafted cellulose nanocrystal (CNC) with lactic acid (LA) oligomer using one-pot synthesis to reinforce poly(L-lactic acid) (PLLA). Structural and mechanical analysis of the PLLA composites was done using X-ray diffraction (XRD), FTIR, ^{13}C-NMR (nuclear magnetic resonance spectroscopy), and tensile test. Corroborating results of the structural and mechanical analysis confirmed the grafting technique to be beneficial in improving the interfacial compatibility of CNC with PLLA. Based on the biological and physicochemical properties, Wang et al. considered PLLA-CNC-*g*-LA composite to be promising for packaging applications [18].

Pasarvi et al. fabricated PLA/PCL/zinc oxide nanocomposites with reduced cytotoxicity, with increased antibacterial activity with the use of thymol for antibacterial wound healing film with 5 wt% of zinc oxide [19].

Ren et al. fabricated an electrospun nanofiber mat containing PLA and chitosan for wound healing applications, that is, to develop a disinfectant wound dressing. In this study, Ren et al. fabricated multifunctional nanofiber mats using electrospinning combining chitosan derivations and stereocomplex crystallite (SC); the SC mats were prepared by warming the electrospun combining chitosan derivations mats at 80°C for 60 min. SC mats showed improvement in the thermal and mechanical properties of the nanofiber mats. Ren et al. claimed that SC mats also showed wound repairability of 100% within 15 days [20]. Similar work on stereocomplex membranes was done by Chuan et al., in which they develop materials with excellent material properties and stem cell affinity for bone repair in the field of bone tissue engineering using stereocomplex PLA composite nanofiber membranes with highly dispersed HAP [21].

Pathania et al. fabricated carboxymethyl cellulose-cl-poly(lactic acid-*co*-itaconic acid)/zinc oxide nanocomposites by grafting lactic acid and itaconic acid onto carboxymethyl cellulose using potassium persulphate and *N,N'*-methylene-bisacrylamide as initiators in a microwave at optimized temperature and pressure. The fabricated nanocomposites proved to be successful for drug delivery application, dye degradation (Congo red), and anticancer behavior [22].

Moura et al. studied the synergistic effects of adding Bioglass and CNTs on PLA porous membranes for guided bone regeneration. Bioglass was added to impart osteoinduction and osteoconduction. CNTs were added to impart antibacterial activity to fabricated PLA membranes. The solvent cast method was used to fabricate the PLA membranes [23].

Almansoori et al. explored tantalum PLA layered composites as a nerve conduit for peripheral nerve regeneration in Sprague Dawley rats. PLA was used as an outer layer and tantalum as an inner layer in the fabrication of a nerve conduit. PLA was electrospun whereas inner coating on PLA porous membrane was done using DC magnetic sputtering [24].

Thummarungsan et al. utilized MWCNTs and dibutyl phthalate in tuning rigidity and electro-responsive properties of PLA composites prepared using solvent casting. Results confirmed that 0.1% v/v MWCNTs in PLA composites showed the highest positive storage modulus sensitivity of 1.95 at an electric field strength of 1.5 kV/mm [25].

Yang et al. synthesized PLA-PCL-nano lignin composites films using ring-opening polymerization to obtain flexible PLA composites resistant to UV. The mechanical test confirmed 6 wt% of nano lignin optimum in contributing to the toughness of PLA films. 12 wt% of nano lignin imparted PLA films to shield 100% of UVB and UVC irradiation. Analyzing the result, the authors proposed PLA-PCL-nano lignin composites as a candidate for sunshade and food packaging applications [26].

Inphonlek et al. fabricated porous scaffolds based on chitosan-xanthan gum incorporated in situ with PLA particles using emulsification-solvent evaporation, followed by lyophilization. Xanthan gum improved the physical stability and PLA particles improved the modulus of chitosan-based scaffolds. Authors proposed that chitosan-xanthan gum-PLA particle composites be a potential candidate for capturing and separating anionic compounds [27].

Silva et al. incorporated lignin into PLA to accelerate the biodegradation of PLA. A melt blender was used to incorporate lignin into PLA. 10 wt% of lignin in PLA increased the degradation rate of PLA and showed similar properties to that of neat PLA [28].

Lemos et al. fabricated electroactive nanofiber mats based on PLA-poly(ortho-ethoxyaniline) (POEA) blends for biological applications. PLA-POEA blends were fabricated into porous mats using the electrospinning technique; the homogeneity of POEA in PLA rendered redox reversibility to PLA-POEA blends, which imparted biocompatibility and noncytotoxic features making it a successful candidate for the biological application [29].

Perić et al. fabricated PLA-nano fibrillated cellulose composite using a twin-screw extruder followed by injection molding. Mechanical and thermal properties of fabricated PLA composites were determined – 5 wt% of nano fibrillated cellulose in PLA improved crystallization behavior of PLA, thus improving the elastic modulus of PLA composites as compared to neat PLA [30].

Samadi et al. studied mechanical, rheological, and anaerobic biodegradation of PLA-*co*-poly(glycolic acid) (PGA) (PLGA) blend in comparison to neat PLA. Glass transition temperature (T_g) of PLGA was lower than that of PLA. Single T_g confirmed the miscibility of blends. In comparison to PLA, the anaerobic biodegradation rate of PLGA was increased with increasing PLGA content and decreasing molecular weight of PLGA, and also tensile strength and tensile modulus decreased for PLGA in comparison to PLA. Zero scale viscosity of PLGA was in direct relation with the PLGA content and exhibited reduced sensitivity to shear rate. Authors proposed that this (anaerobic digestion) technique be used at places such as universities, hospitals, and stadiums where the waste stream consists of plastic products associated with paper generating biogas as a fuel [31].

Louisy et al. reported the synthesis of a novel hybrid PLA by use of titanium oxo-cluster acting as both a catalyst and a cross-linking agent via reactive emulsion polymerization. 90% conversion in 30 min was obtained [32].

Suwanamornlert et al. fabricated PLA/poly(butylene-succinate-*co*-adipate) (PBSA)-thymol films for active packaging of bread. The PLA blend films were fabricated via the blown film extrusion technique. The addition of PBSA and thymol to PLA decreased the T_g, with a decrease in mechanical properties in comparison to neat PLA film. Permeability coefficients (water vapor, oxygen, and carbon dioxide) for PLA-PBSA-thymol film were lower as compared to neat PLA film. Increased thymol in PLA-PBSA delayed the mold growth on

PLA-PBSA film in comparison to the rate of mold growth on biaxially oriented polypropylene (BOPP) film and PLA-PBSA film. In-package carbon dioxide concentration was observed highest for BOPP film > PLA-PBSA-thymol film [33].

Studies on miscibility, thermal properties, degradation patterns, and toughening behavior of poly(ethylene-butyl acrylate-glycidyl methacrylate) (PTW) (toughening agent) on PLA were done by Zhao et al. Chemical interactions between an epoxy group of PTW and end groups of PLA increased the compatibility level of PLA-PTW blends; impact strength of PLA-PTW blends increased with increase in PTW content in PLA-PTW blends; addition of PTW to PLA-PTW blends accelerated the biodegradation of PLA-PTW blends. PLA-PTW blends were fabricated using a melt mixer [34].

Orue et al. fabricated PLA composites using alkali treated walnut shell flour and plasticized linseed oil (plasticizer). Alkali-treated walnut shell flour in plasticized PLA improved the tensile strength by 50% in comparison to plasticized PLA with untreated walnut shell flour. The tensile strength of alkali-treated walnut shell flour was similar to unfilled plasticized PLA [35].

Geng et al. fabricated ultra-strong PLA composites using PLA and surface-modified cellulose nanocrystals using liquid assisted extrusion and solid-state drawing. A strength of 353 MPa and a toughness of 107 MJ/m^3 are achieved, which are high as compared to many other thermoplastics. Sliding of the PLA crystallites is the main factor for an increase in the toughness of PLA composites. Authors reported that the PLA composites possess high T_g and exhibit high responsive birefringence behavior that make these composites well suited for structural and optical strain sensing areas [36].

6.2 Processing

The wide usage of polymeric materials in engineering is largely due to their valuable mechanical properties. Fracture is a rupture of the bonds between elements of a body (atoms, molecules, or ions) resulting in breakage or cleavage of the specimen into parts. The resistance of a material to fracture is called strength or mechanical strength. Since the mechanical properties of polymers largely depend on their structure, it is necessary to create a structure ensuring an optimal set of mechanical properties that do not vary with time. The structure of the polymer is established during processing. Processing not only imparts a certain shape to the material but also plays an important role in the creation and determination of its structure, that is, microstructures [10]. Structures are often conceived in the melts or solutions from which the polymers are fabricated, and hence the processing method selected for polymer processing depends on the conversion of polymer to molded parts, films, and fibers. Table 6.1 gives information on the latest research work related to the processing of PLA, PLA blends, and PLA composites. Figures 6.3 and 6.4 show the general routes to fabricate PLA composites. PLA can be processed on the standard equipment with slight modification with a consideration given to the form of the product, that is, molded part, film, or fiber. Drying before processing of PLA is an essential step, as moisture in PLA effects the physical and mechanical properties of the PLA and PLA composites by disrupting the molecular weight of the PLA. An extruder is the most important equipment for continuous melt processing of PLA and is associated with most of the forming machines, that is, injection molding, film blowing, and melt spinning. Products with complex shapes and requiring high precision in dimensions are fabricated using injection molding [2].

Table 6.1 Data on general processing technique (GPT) followed by specific processing technologies (SPT) used to fabricate PLA composites

In blend with	Particulate filler/s	GPT → SPT	Ref.
Polycaprolactone	–	Solution mixing – jet spinning	[4]
Zein/tetracycline hydrochloride	Gum tragacanth	Solution mixing – spinning	[5]
Poly(butylene adipate-*co*-terephthalate)	–	Melt mixing – compression molding	[6]
Poly(ethylene oxide)	Graphene nanoplatelets	Melt mixing – compression molding	[7]
Ethylene vinyl acetate	–	Melt mixing – compression molding	[8]

In blend with	Particulate filler/s	GPT → SPT	Ref.
Natural rubber	–	Melt mixing – compression molding	[9]
Poly(ε-caprolactone)	Multiwalled carbon nanotubes	Melt mixing – compression molding	[11]
–	Lithium iron phosphate	Solution mixing – electrospinning	[12]
–	Carbon black	Solution mixing – casting	[37, 10]
–	Carbon black	Melt mixing – compression molding	[37, 10]
Gelatin	–	Solution mixing – foaming	[13]
Poly(vinyl alcohol)	Pulverized PLA	Solution mixing – freeze-drying	[15]
–	Silver nanoparticles	Solution mixing	[14]
–	Hydroxyapatite nanoparticles	Solution mixing – laser powder bed fusion	[16]
–	Hydroxyapatite	Melt mixing – 3D printing	[17]
L-lactic acid oligomer	Nanocellulose crystals	Solution mixing – drying – compression molding	[18]
Poly(ε-caprolactone)	Zinc oxide nanoparticles	Melt mixing – compression molding	[19]
Poly(ε-caprolactone)	Kaolin	Melt mixing – compression molding	[38]
Poly(D-lactic acid)	Microcrystalline cellulose	Solution mixing – casting	[39]
–	Quaternized chitosan	Solution mixing – electrospinning	[20]
–	Hydroxyapatite	Solution mixing – electrospinning	[21]

(*Continued*)

Table 6.1 (*Continued*)

In blend with	Particulate filler/s	GPT → SPT	Ref.
–	Zinc oxide – silver nanoparticles	Solution mixing – coprecipitation	[22]
–	Bioglass – multiwalled carbon nanotubes	Solution mixing – casting	[23]
Dibutyl phthalate	Multiwalled carbon nanotubes	Solution mixing – casting	[25]
Poly(ε-caprolactone)	Nano lignin	Solution mixing – casting	[26]
–	Chitosan and xanthan gum	Solution mixing – freeze-drying	[27]
–	Lignin	Melt mixing – compression molding	[28]
Poly(ortho-ethoxy aniline)	–	Solvent mixing – electrospinning	[29]
–	Nano fibrillated cellulose	Melt mixing – compression molding	[30]
Poly(glycolic acid)	–	Melt mixing – compression molding	[31]
Poly(butylene-succinate-*co*-adipate)	–	Melt mixing – blow film single screw extruder	[33]
Poly(ethylene-butyl acrylate-glycidyl methacrylate)	–	Melt mixing – compression molding	[34]
Acrylonitrile-butadiene-styrene	Tapioca cassava starch	Melt mixing – compression molding	[40]
–	Walnut shells	Melt mixing – compression molding	[35]
–	Cellulose nanocrystals	Melt mixing – liquid assisted feeding – solid-state drawing – compression molding	[36]

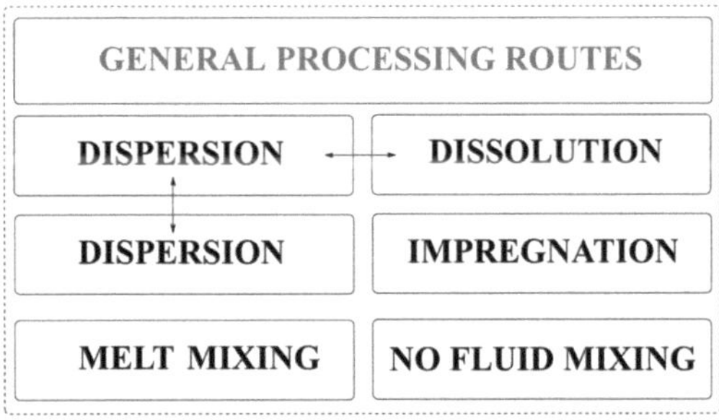

Figure 6.3 General routes for processing poly(lactic acid) and poly(lactic acid) composites (PLAC).

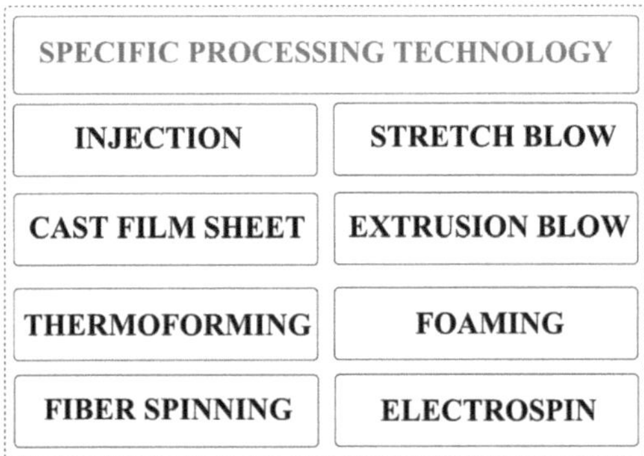

Figure 6.4 Specific processing technologies for processing poly(lactic acid) and poly(lactic acid) composites (PLAC).

Stretch blow molding is used to prepare PLA bottles that are supplied to the food and beverage industry. Biaxial oriented PLA bottles with enhanced physical and barrier properties are fabricated using an injection stretch blow molding technology in comparison to PLA bottles prepared by injection molding using amorphous PLA. Molten PLA is extruded through a sheet die of an extruder and is quenched on polished chrome horizontal rollers that are cooled with circulating water. Molten PLA is extruded through an annular die to form a tube; the air is blown through the head of the annular die, which inflates the tube into a thin tubular bubble and cooled. Forming packaging containers are fabricated using a thermoforming technology, wherein the PLA sheet is softened by heating and forced against the mold using mechanical or pneumatic controls, cooled, detached from the mold, and trimmed. Foaming technology is used to fabricate products used in medical applications, wherein a blowing agent is dissolved in a PLA matrix; further inducing thermodynamic instability in the structure reduces the solubility ability of the blowing agent that forms bubbles, which are further stabilized by vitrification. Dry or melt spinning process is used to fabricate PLA fibers, and usually a melt spinning process is preferred commercially; PLA is heated and extruded out of spinnerets in the first step and cooled, followed by a stretch, that is, hot drying to achieve fiber orientation. Wet spinning is almost similar to melt spinning wherein PLA solution is spun into a coagulation bath containing coagulating solution, which solidifies the PLA filament. Fibers with a much smaller diameter can be produced using electrospinning, with PLA solution – similar to dry spinning, however, an electrostatic force is used to form a filament [2].

6.3 Properties

Thermal, chemical, and importantly mechanical characterizations and aspects of PLA and PLA composites are important concerning their effect on properties of PLA and PLA composites. The ratio of "L" and "D" stereochemistry in PLA has specific effects on the properties of PLA. L:D ratio consequently affects the crystallinity in PLA, which further affects the melting temperature and glass transition temperature. At temperature >200°C, PLA starts degrading into lactide and oxide gases. Studies prove that PLA has low solubility toward a wide range of solvents and liquids such as water, alcohol, and paraffin; in addition to low solubility, PLA also possesses good barrier properties, which makes PLA a preferable candidate for food packaging. High stereochemical pure PLA possesses high tensile strength and modulus but lacks impact strength, in contrast to a copolymer of "L" and "D" forms, an amorphous PLA that has poor mechanical properties [1].

6.4 Applications

PLA is a biodegradable polymer that has a variety of applications as listed in Fig. 6.5. PLA finds its applications in the medical

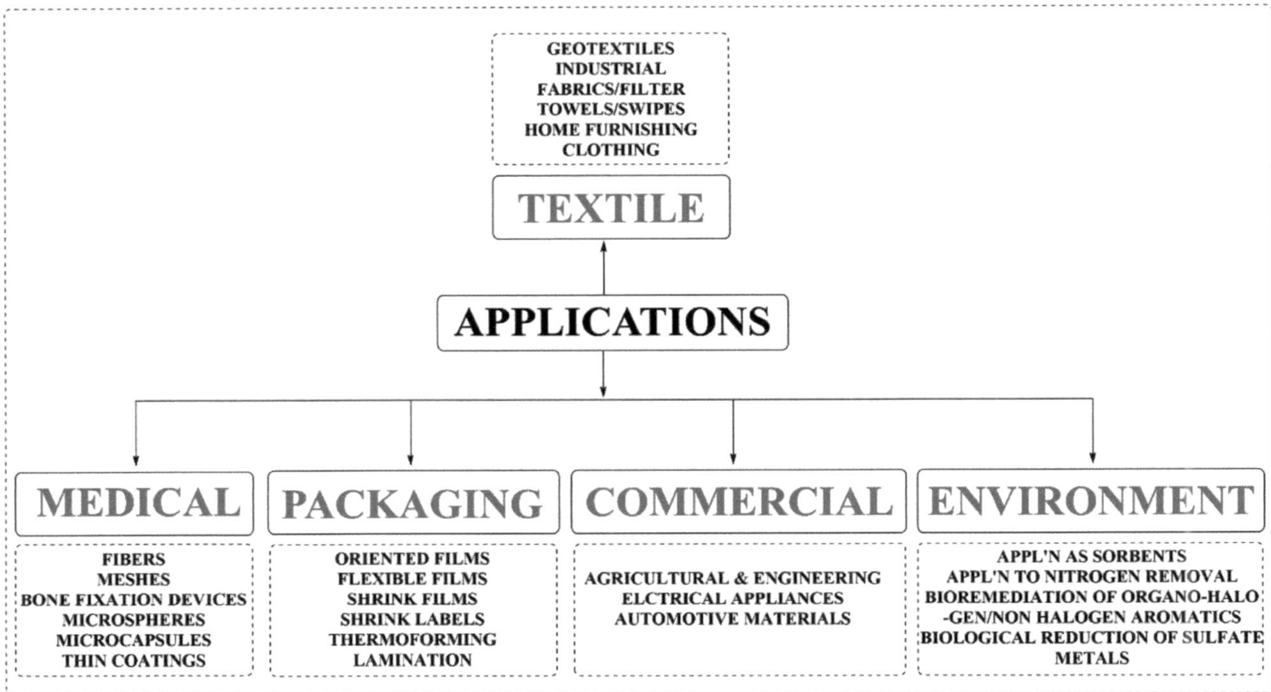

Figure 6.5 Applications of PLAC.

and pharmaceutical industry, owing to its biocompatibility and biodegradability. The application of PLA was limited due to its high cost involved in the synthesis of PLA in the laboratory [1]. Owing to development in the synthesis of PLA on a large scale, PLA finds its suitability in the majority of the applications. PLA in itself is a brittle polymer, with good strength, but its inability to deformation on the application of stress limits its applications in fabricating a structural component; hence, many researchers as mentioned above developed processing techniques to increase the deformation without compromising on the strength [1, 41]. Addition of fillers, low molecular weight oligomers, and polymers with specific processing techniques imparted specific property to PLA for a specific application [4, 5].

6.5 Developments

PLA has gained tremendous attention due to increasing environmental pressure on global warming and plastic pollution. The global market for PLA demand doubles every 3 or 4 years as estimated by Jem's law [3]. PLA is more expensive and has less mechanical properties in comparison to traditional petroleum-based plastics. The recent compounding efforts and the commercialization of D(−) lactic acid and its polymer PDLA have the potential to improve the mechanical and thermal properties of PLA composites and PLA (by forming stereocomplex PLA), respectively. The development of novel production technology and the advent of government regulations are the key drivers for the global transition toward bioplastics. Restriction on the use of petroleum-based plastics and facilitating the use of biodegradable plastics have been released by multiple governments across the globe. PGA can be combined with PLA using new production and compounding technology to have a sustainable and environmentally friendly plastic industry. Special attention should be given to single used products with fast degradation at room temperature or in the environment [3].

6.6 Conclusions

PLA is a biodegradable and biocompatible polymer that has found use in a wide range of applications. The development of PLA in a wide range of industrial applications is predicted to grow tremendously in the future, thereby making the cost of PLA economical as commodity plastics with the added advantage of being caring to the environment. An environment-friendly approach and a substitute for petrochemical-based polymers make PLA "Green" polymer. Biodegradability of PLA, PLA blends, and PLA composites should be considered while compounding a PLA (PLA blends/PLA composites) for mechanical properties.

Acknowledgment

The first author is thankful to CSIR and DST – South Africa for providing the CSIR – DST Inter-Programme Bursary Award in the area of Composites for the years 2016, 2017, and 2018. The authors also extend thanks to Durban University of Technology, South Africa, and Mahatma Gandhi University, Kerala, India for their support to carry out this work. The authors also wish to thank all the authors whose work has been referred to while writing this chapter.

References

1. L.T. Sin, A.R. Rahmat, W.W. Rahman, Poly(lactic acid): PLA Biopolymer Technology and Applications, 1st ed., Elsevier Ltd, Great Britain, 2012.
2. L.T. Lim, R. Auras, M. Rubino, Processing technologies for poly(lactic acid), *Prog. Polym. Sci.* 33 (2008) 820–852. doi:10.1016/j.progpolymsci.2008.05.004.
3. K.J. Jem, B. Tan, The development and challenges of poly (lactic acid) and poly (glycolic acid), *Adv. Ind. Eng. Polym. Res.* 3 (2020) 60–70. doi:10.1016/j.aiepr.2020.01.002.
4. T. Hongthipwaree, P. Sriamornsak, M. Seadan, S. Suttiruengwong, Effect of cosolvent on properties of non-woven porous neomycin-loaded poly(lactic acid)/polycaprolactone fibers, *Mater. Today Sustain.* 10 (2020) 100051. doi:10.1016/j.mtsust.2020.100051.
5. M. Ghorbani, F. Mahmoodzadeh, L. Yavari Maroufi, P. Nezhad-Mokhtari, Electrospun tetracycline hydrochloride loaded zein/gum tragacanth/poly lactic acid nanofibers for biomedical application, *Int. J. Biol. Macromol.* 165 (2020) 1312–1322. doi:10.1016/j.ijbiomac.2020.09.225.

6. Y. Zhao, B. Zhao, B. Wei, Y. Wei, J. Yao, H. Zhang, X. Chen, Z. Shao, Enhanced compatibility between poly(lactic acid) and poly (butylene adipate-co-terephthalate) by incorporation of N-halamine epoxy precursor, *Int. J. Biol. Macromol.* 165 (2020) 460–471. doi:10.1016/j.ijbiomac.2020.09.142.
7. Q. Zeng, Z. Du, C. Qin, Y. Wang, C. Liu, C. Shen, Enhanced thermal, mechanical and electromagnetic interference shielding properties of graphene nanoplatelets-reinforced poly(lactic acid)/poly(ethylene oxide) nanocomposites, *Mater. Today Commun.* 25 (2020) 1–9. doi:10.1016/j.mtcomm.2020.101632.
8. S. Moradi, J.K. Yeganeh, Highly toughened poly(lactic acid) (PLA) prepared through melt blending with ethylene-co-vinyl acetate (EVA) copolymer and simultaneous addition of hydrophilic silica nanoparticles and block copolymer compatibilizer, *Polym. Test.* 91 (2020). doi:10.1016/j.polymertesting.2020.106735.
9. W. Tessanan, R. Chanthateyanonth, M. Yamaguchi, P. Phinyocheep, Improvement of mechanical and impact performance of poly(lactic acid) by renewable modified natural rubber, *J. Clean. Prod.* 276 (2020) 123800. doi:10.1016/j.jclepro.2020.123800.
10. A. V. Rane, K. Kanny, A. Mathew, T.P. Mohan, S. Thomas, Comparative analysis of processing techniques' effect on the strength of carbon black (N220) filled poly (lactic acid) composites, *Strength Mater.* 51 (2019) 476–489. doi:10.1007/s11223-019-00093-6.
11. X.H. Tang, Y. Tang, Y. Wang, Y.X. Weng, M. Wang, Interfacial metallization in segregated poly (lactic acid)/poly (ε-caprolactone)/multi-walled carbon nanotubes composites for enhancing electromagnetic interference shielding, *Compos. Part A Appl. Sci. Manuf.* 139 (2020) 106116. doi:10.1016/j.compositesa.2020.106116.
12. J.C. Barbosa, A. Reizabal, D.M. Correia, A. Fidalgo-Marijuan, R. Gonçalves, M.M. Silva, S. Lanceros-Mendez, C.M. Costa, Lithium-ion battery separator membranes based on poly(L-lactic acid) biopolymer, *Mater. Today Energy.* 18 (2020) 100494. doi:10.1016/j.mtener.2020.100494.
13. R. Rarima, G. Unnikrishnan, Poly(lactic acid)/gelatin foams by non-solvent induced phase separation for biomedical applications, *Polym. Degrad. Stab.* 177 (2020) 109187. doi:10.1016/j.polymdegradstab.2020.109187.
14. R. Rarima, G. Unnikrishnan, Porous poly(lactic acid)/nano-silver composite membranes for catalytic reduction of 4-nitrophenol, *Mater. Chem. Phys.* 241 (2020) 122389. doi:10.1016/j.matchemphys.2019.122389.

15. K. Chronaki, D.M. Korres, C.D. Papaspyrides, S. Vouyiouka, Poly(lactic acid) microcapsules: Tailoring properties via solid state polymerization, *Polym. Degrad. Stab.* 179 (2020) 109283. doi:10.1016/j.polymdegradstab.2020.109283.

16. D. Yan, B. Zeng, Y. Han, H. Dai, J. Liu, Y. Sun, F. Li, Preparation and laser powder bed fusion of composite microspheres consisting of poly(lactic acid) and nano-hydroxyapatite, *Addit. Manuf.* 34 (2020) 101305. doi:10.1016/j.addma.2020.101305.

17. G.E. Dubinenko, A.L. Zinoviev, E.N. Bolbasov, V.T. Novikov, S.I. Tverdokhlebov, Preparation of poly(L-lactic acid)/hydroxyapatite composite scaffolds by fused deposit modeling 3D printing, *Mater. Today Proc.* 22 (2020) 228–234. doi:10.1016/j.matpr.2019.08.092.

18. K. Wang, J. Lu, R. Tusiime, Y. Yang, F. Fan, H. Zhang, B. Ma, Properties of poly (L-lactic acid) reinforced by L-lactic acid grafted nanocellulose crystal, *Int. J. Biol. Macromol.* 156 (2020) 314–320. doi:10.1016/j.ijbiomac.2020.04.025.

19. S. Kazemi-Pasarvi, N. Golshan Ebrahimi, D. Shahrampour, Z. Arab-Bafrani, Reducing cytotoxicity of poly (lactic acid)-based/zinc oxide nanocomposites while boosting their antibacterial activities by thymol for biomedical applications, *Int. J. Biol. Macromol.* 164 (2020) 4556–4565. doi:10.1016/j.ijbiomac.2020.09.069.

20. Y. Ren, L. Huang, Y. Wang, L. Mei, R. Fan, M. He, C. Wang, A. Tong, H. Chen, G. Guo, Stereocomplexed electrospun nanofibers containing poly (lactic acid) modified quaternized chitosan for wound healing, *Carbohydr. Polym.* 247 (2020) 116754. doi:10.1016/j.carbpol.2020.116754.

21. D. Chuan, R. Fan, Y. Wang, Y. Ren, C. Wang, Y. Du, L. Zhou, J. Yu, Y. Gu, H. Chen, G. Guo, Stereocomplex poly(lactic acid)-based composite nanofiber membranes with highly dispersed hydroxyapatite for potential bone tissue engineering, *Compos. Sci. Technol.* 192 (2020) 108107. doi:10.1016/j.compscitech.2020.108107.

22. D. Pathania, S. Sood, A.K. Saini, S. Kumari, S. Agarwal, V.K. Gupta, Studies on anticancerious and photocatalytic activity of carboxymethyl cellulose-cl-poly(lactic acid-co-itaconic acid)/ZnO-Ag nanocomposite, *Arab. J. Chem.* 13 (2020) 6966–6976. doi:10.1016/j.arabjc.2020.07.001.

23. N.K. de Moura, E.F. Martins, R.L.M.S. Oliveira, I.A.W. de Brito Siqueira, J.P.B. Machado, E. Esposito, S.S. Amaral, L.M.R. de Vasconcellos, F.R. Passador, E. de Sousa Trichês, Synergistic effect of adding bioglass and carbon nanotubes on poly (lactic acid) porous membranes for guided bone regeneration, *Mater. Sci. Eng. C.* 117 (2020) 111327. doi:10.1016/j.msec.2020.111327.

24. A.A. Almansoori, C. Hwang, S.H. Lee, B. Kim, H.E. Kim, J.H. Lee, Tantalum – Poly (L-lactic acid) nerve conduit for peripheral nerve regeneration, *Neurosci. Lett.* 731 (2020) 135049. doi:10.1016/j.neulet.2020.135049.
25. N. Thummarungsan, D. Pattavarakorn, A. Sirivat, Tuning rigidity and negative electrostriction of multi-walled carbon nanotube filled poly(lactic acid), *Polymer (Guildf)* 196 (2020) 122488. doi:10.1016/j.polymer.2020.122488.
26. W. Yang, G. Qi, H. Ding, P. Xu, W. Dong, X. Zhu, T. Zheng, P. Ma, Biodegradable poly (lactic acid)-poly (ε-caprolactone)-nanolignin composite films with excellent flexibility and UV barrier performance, *Compos. Commun.* 22 (2020) 100497. doi:10.1016/j.coco.2020.100497.
27. S. Inphonlek, N. Niamsiri, P. Sunintaboon, C. Sirisinha, Chitosan/xanthan gum porous scaffolds incorporated with in-situ-formed poly(lactic acid) particles: Their fabrication and ability to adsorb anionic compounds, *Colloids Surfaces A Physicochem. Eng. Asp.* 603 (2020) 125263. doi:10.1016/j.colsurfa.2020.125263.
28. T.F. da Silva, F. Menezes, L.S. Montagna, A.P. Lemes, F.R. Passador, Effect of lignin as accelerator of the biodegradation process of poly(lactic acid)/lignin composites, *Mater. Sci. Eng. B Solid-State Mater. Adv. Technol.* 251 (2019) 114441. doi:10.1016/j.mseb.2019.114441.
29. H.G. de Lemos, L.M.G. da Silva, F.N. Ambrosio, C.B. Lombello, J.C. Moreira, E.C. Venancio, Electroactive nanofibers mats based on poly(L-lactic acid)/poly(ortho-ethoxyaniline) blends for biological applications, *Mater. Sci. Eng. C.* 105 (2019) 110045. doi:10.1016/j.msec.2019.110045.
30. M. Perić, R. Putz, C. Paulik, Influence of nanofibrillated cellulose on the mechanical and thermal properties of poly(lactic acid), *Eur. Polym. J.* 114 (2019) 426–433. doi:10.1016/j.eurpolymj.2019.03.014.
31. K. Samadi, M. Francisco, S. Hegde, C.A. Diaz, T.A. Trabold, E.M. Dell, C.L. Lewis, Mechanical, rheological and anaerobic biodegradation behavior of a poly(lactic acid) blend containing a poly(lactic acid)-co-poly(glycolic acid) copolymer, *Polym. Degrad. Stab.* 170 (2019) 109018. doi:10.1016/j.polymdegradstab.2019.109018.
32. E. Louisy, S. Bellayer, G. Fontaine, L. Rozes, F. Bonnet, Novel hybrid poly(L-lactic acid) from titanium oxo-cluster via reactive extrusion polymerization, *Eur. Polym. J.* 122 (2020) 109238. doi:10.1016/j.eurpolymj.2019.109238.
33. P. Suwanamornlert, N. Kerddonfag, A. Sane, W. Chinsirikul, W. Zhou, V. Chonhenchob, Poly(lactic acid)/poly(butylene-succinate-co-adipate) (PLA/PBSA) blend films containing thymol as alternative to synthetic

preservatives for active packaging of bread, *Food Packag. Shelf Life*. 25 (2020) 100515. doi:10.1016/j.fpsl.2020.100515.

34. J. Zhao, H. Pan, H. Yang, J. Bian, H. Zhang, G. Gao, L. Dong, Study on miscibility, thermal properties, degradation behaviors, and toughening mechanism of poly(lactic acid)/poly (ethylene-butylacrylate-glycidyl methacrylate) blends, *Int. J. Biol. Macromol.* 143 (2020) 443–452. doi:10.1016/j.ijbiomac.2019.11.226.
35. A. Orue, A. Eceiza, A. Arbelaiz, The use of alkali treated walnut shells as filler in plasticized poly(lactic acid) matrix composites, *Ind. Crops Prod.* 145 (2020) 111993. doi:10.1016/j.indcrop.2019.111993.
36. S. Geng, D. Wloch, N. Herrera, K. Oksman, Large-scale manufacturing of ultra-strong, strain-responsive poly(lactic acid)-based nanocomposites reinforced with cellulose nanocrystals, *Compos. Sci. Technol.* 194 (2020) 108144. doi:10.1016/j.compscitech.2020.108144.
37. A.V. Rane, K. Kanny, A. Mathew, M.T. Pandurangan, S. Thomas, Microstructural features affecting mechanical properties: Effect of processing on dispersion of carbon black (N220) nanoparticles reinforcement in poly (lactic acid), *Surf. Interfaces* 18 (2020) 100451. doi:10.1016/j.surfin.2020.100451.
38. X. Ge, M. Chang, W. Jiang, B. Zhang, R. Xing, C. Bulin, Selective location of kaolin and effects of maleic anhydride in kaolin/poly(ε-caprolactone)/poly(lactic acid) composites, *Appl. Clay Sci.* 189 (2020) 105524. doi:10.1016/j.clay.2020.105524.
39. H. Rahaman, S. Hosen, A. Gafur, R. Habib, Small amounts of poly(D-lactic acid) on the properties of poly(L-lactic acid)/microcrystalline cellulose/poly(D-lactic acid) blends, *Results Mater.* 8 (2020) 100125. doi:10.1016/j.rinma.2020.100125.
40. B. Ramanjaneyulu, N. Venkatachalapathi, G. Prasanthi, Testing and characterization of binary and ternary blends with poly (lactic acid), acrylonitrile-butadiene-styrene and tapioca cassava starch powder, *Mater. Today Proc.* 27 (2019) 2183–2186. doi:10.1016/j.matpr.2019.09.092.
41. C. Bastioli, Handbook of Biodegradable Polymers, 1st ed., Rapra Technology, United Kingdom, 2005.

Chapter 7

Green Composites Based on Protein Materials

Sabana Ara Begum,[a,b] Ajay Vasudeo Rane,[c] Deepti Yadav,[d] P. Santhana Gopala Krishnan,[b] Treesa Reji,[e] and Krishnan Kanny,[a]

[a]*Composite Research Group, Department of Mechanical Engineering, Faculty of Engineering and Built Environment, Durban University of Technology, Durban-4000, South Africa*

[b]*Advanced Polymer Design and Development Research Laboratory (APDDRL), School for Advanced Research in Polymers (SARP), Central Institute of Petrochemicals Engineering and Technology (CIPET), Bengaluru-562149, India*

[c]*Materials Research Laboratory, Department of Mechanical Engineering, KMEA Engineering College, Cochin, Kerala, India*

[d]*Department of Biotechnology and Food Science, Durban University of Technology, Durban-4000, South Africa*

[e]*School of Nanoscience and Nanotechnology, Mahatma Gandhi University, Kottayam-686 560, Kerala, India*

avr.me@kmeacollege.ac.in

7.1 Introduction to Proteins and Protein Composites

At the beginning of the 20th century, proteins were used to produce edible packaging and materials. Later the invention of "synthetic

Green Micro- and Nanocomposites
Edited by Sabu Thomas, Abitha V. K., and Hanna J. Maria

ISBN 978-981-4968-79-9 (Hardcover), 978-1-003-42756-8 (eBook)
www.jennystanford.com

polymers" in 1960 led to the abandonment of protein materials for nearly 30 years, but an increasing number of research programs at academic and industrial levels since 1980 resulted in interest in the usage of field crops for renewable and biodegradable materials for nonfood applications. Proteins are generally composed of a linear combination of the 20 naturally occurring L-α-amino acids (see Fig. 7.1) and are thereby polyamides [1]. Amino acids are classified

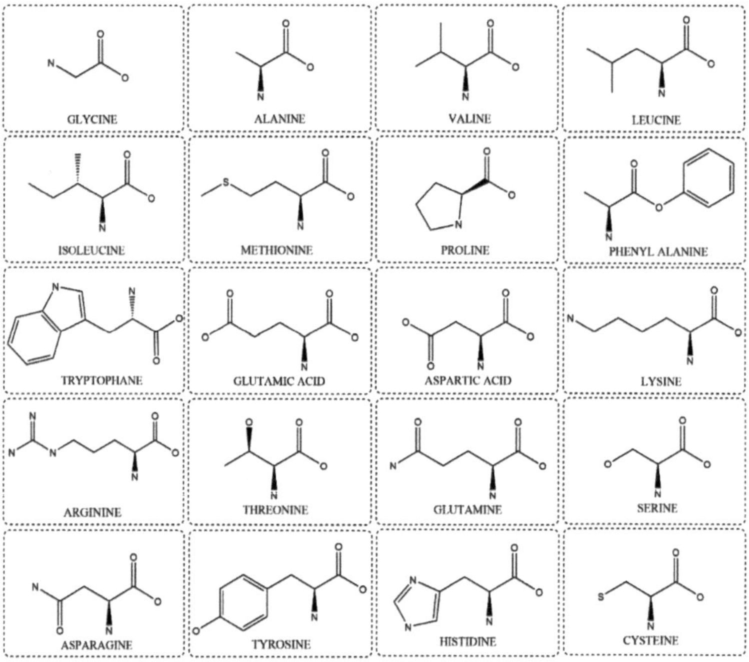

Figure 7.1 Chemical structures of amino acids.

on basis of the chemical groups that could interact as nonpolar amino acids (Glycine ↔ Tryptophane), ionized polar amino acids (Glutamic acid ↔ Arginine), nonionized polar amino acids (Threonine ↔ Histidine), and amino acids able to form –SS– bond (Cysteine) [1]. The polymerization of the amino acids occurs through the formation of an amide linkage between the carboxyl group of the given amino acid and the amino group of the next amino acid, with the resultant elimination of a molecule of water. This amide linkage is more commonly called a "peptide bond" and proteins are "polypeptides." The molecular basis for the enormously diverse functions of the proteins is the significant diversity of their building blocks. The chemical nature of the 20 natural amino acids is extremely versatile. The side chain of the amino acids can be either negatively or positively charged, polar, aliphatic (branched or unbranched), or aromatic (substituted or non-substituted). It may include various functional groups such as thiol, amine, hydroxyl, carboxyl, phenyl, and amide groups. Furthermore, the properties of this biopolymer are a result of the amino acid sequence, that is, the linear arrangement of the building blocks, rather than the composition of the protein. Two proteins with the very same amino acid composition may be completely different in their folded state and molecular properties [2].

Most of the proteins found are heteropolymers, which consist of different types of amino acids with specific sequences and structures [2]. Considering the complexity of proteins and their variable fractions, and understanding their structural details can be helpful to develop composite materials with specific functional property/ies.

Classification of proteins based on their chemical composition, function, shape, size, and source are shown in Fig. 7.2. For a better understanding of protein materials used in the fabrication of composite materials, we shall consider classification based on the source. Figure 7.3 enlists the proteins from plant and animal sources. Protein may not be able to solve all material problems such as high temperature, chemical resistance, environmental stability, and cost and hence will make dependency on petroleum-based polymers (synthetic polymers). However, the biodegradability, production sources, and routes make them suitable for the future. Further, the ability to use composite fabricated using protein and synthetic polymers appears favorable to tailor-make "protein composites." Protein composites can be classified into three major categories: protein in protein, protein in the polymer (here polymer refers to biobased and petroleum-based polymers), and polymer in protein as shown in Fig. 7.4. In the later part of the chapter, we shall discuss the processing, properties, and applications of protein composites.

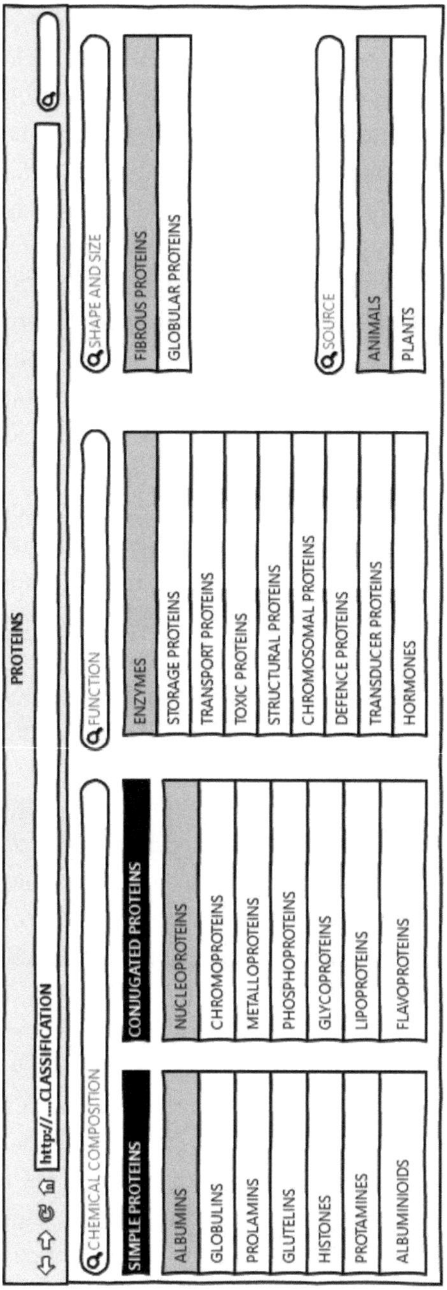

Figure 7.2 Classification of proteins.

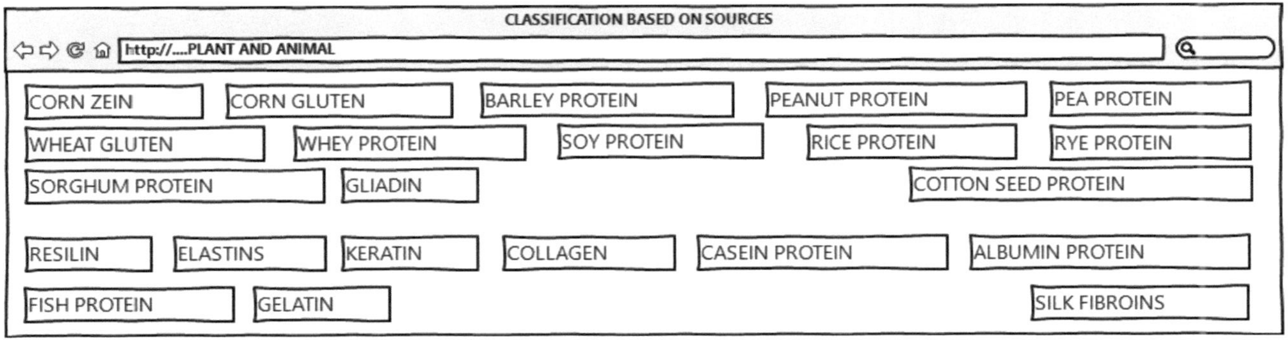

Figure 7.3 Classification of proteins based on sources (i.e., plants and animals).

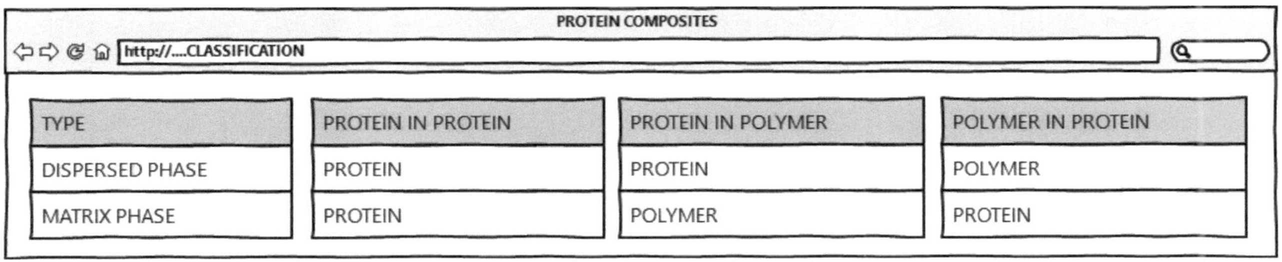

Figure 7.4 Classification of protein composites.

7.2 Processing of Protein Composites

Figure 7.5 shows the general description of the wet and dry processes used for the fabrication of protein composites.

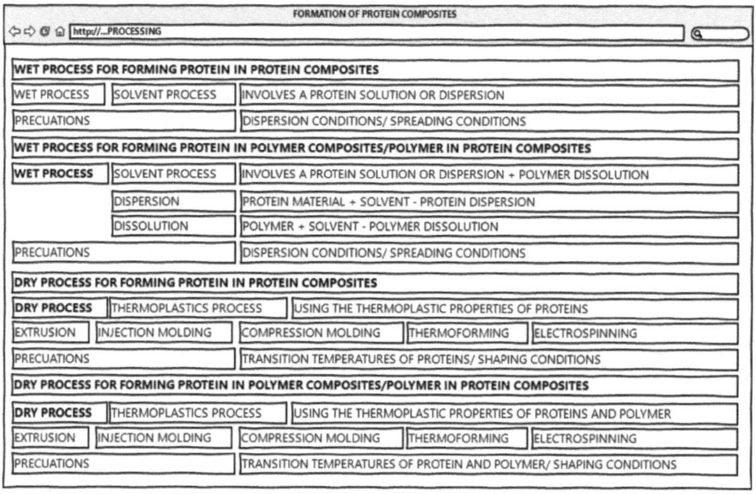

Figure 7.5 Technological approaches to fabricating protein material based green composites.

7.2.1 Wet Process/Solvent Process

The wet process/solvent process is a complete laboratory controlled process and rarely practiced on an industrial scale. The solvent process for protein in protein type of protein composites involves the preparation of homogeneous protein dispersion and further casting the dispersion uniformly over a flat surface to form a film of the required dimension. For protein in the polymer type of protein composites, a homogeneous dispersion of protein and a polymer dissolution is prepared and suspensions of protein in polymer dissolutions are prepared and cast on a plain surface to form a film or sheet of required dimension; a similar approach of processing can be adopted for polymer in protein type of protein composites. It was confirmed by Osborne that the solubility of proteins is highly variable; there are no specific solubilization conditions for casting a film, and generally the nature of different intermolecular

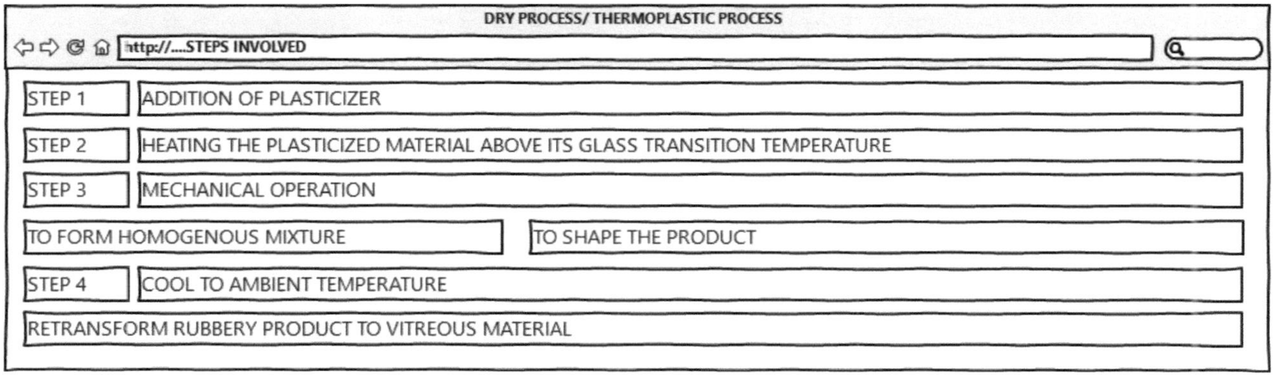

Figure 7.6 Different steps involved in the fabrication of a protein composite using dry process/thermoplastic process.

interactions should be known. Hence, the properties of the protein composites are dependent on the concentration of each protein (in the "protein in protein" type of protein composites) and the concentration of polymer and protein (in "protein in polymer" and "polymer" in "protein" types of protein composites); apart from the concentration of individual constituents, pH, the polarity of solvent system, solution drying time, temperature, and rate should also be taken in account [1, 3].

7.2.2 Dry Process/Thermoplastic Process

The dry process/thermoplastic process is practiced at laboratory as well as industrial scale. The thermoplastic nature of the proteins (in "protein in protein" type of protein composite) and of proteins and polymers (in "protein in polymer" and "polymer in protein" type of protein composites) are utilized by thermal or thermomechanical process under controlled conditions. Thermoplastic nature relates closely to the glass transition temperature (T_g) of proteins and polymers used in the fabrication of protein composites. Molecular weight, chain rigidity, size and polarity of the lateral groups, intermolecular bonds, crystalline zones, plasticizer type, and concentration effect the T_g of proteins and polymers.

The plasticizer is the only material factor that can be easily varied for effective processing of protein composites (after setting machine parameters). Hence, the addition of a plasticizer is the first step during the fabrication of a protein composite. Figure 7.6 shows the different steps involved in the fabrication of protein composites by using the dry process [1].

7.3 Properties of Protein Composites

Macroscopic properties depend partially on the system-stabilizing interactions (i.e., intermolecular–intramolecular interactions). Most of the protein composites are used for barrier applications, with optimum mechanical properties. Mechanical properties in protein composites can be partly related to the distribution of intermolecular and intramolecular interactions, that is, binding energy. An increase in binding energy between the protein components will make the

PROTEIN COMPOSITES

http://....PROPERTIES

PROPERTIES	PROTEIN IN PROTEIN	PROTEIN IN POLYMER	POLYMER IN PROTEIN
MECHANICAL	x	+	x
WATER PERMEABILITY	+	x	+
GAS BARRIER (LOW HUMIDITY)	+	+	+

PROPERTIES	PROTEIN IN PROTEIN	PROTEIN IN POLYMER	POLYMER IN PROTEIN
MECHANICAL	<<	1	>>
WATER PERMEABILITY	>>	1	<<
GAS BARRIER (LOW HUMIDITY)	<<	1	>>

Figure 7.7 Properties of protein composites.

protein composite relatively elastic and a decrease in binding energy will make the protein composite ductile. Barrier properties of protein composites depend on the proportion and distribution of nonpolar amino acids relative to polar amino acids. Protein composites have high water vapor permeability in comparison to synthetic polymer composite materials and therefore are interesting for applications that need "breathe" (i.e., packaging of the fresh product, films for agriculture, and cosmetic applications). Gas barrier properties of protein composites are interesting in low relative humid conditions. Aroma barrier properties of protein composites seem attractive [1]. The general description of the comparison of barrier properties and mechanical properties for protein composites is represented in Fig. 7.7.

7.4 Applications of Protein Composites

The complex chemical composition of proteins can be tuned and altered depending on the processing technique selected for a suitable application. Figure 7.8 enlists the use of protein composites in specific applications [1].

7.5 Selected Works on Protein Composites

Keratin is one of the most generous structural proteins, and in animals together with collagen, it is one of the most essential biopolymers. Keratin is a wide family of animal proteins with a broad variety of properties and morphologies and is the major element of feathers, hair, and horns [4]. Keratins constitute the highest subgroup of intermediate filament (IF) proteins and create the bulk of cytoplasmic epithelial and epidermal appendageal structures [5, 6]. Keratins are a group of fibrous structural proteins rich in cysteine and have higher mechanical properties due to the major quantity of disulfide bonds. Based on the hydrogen bonding of the primary protein structure, keratins are categorized into two different groups, namely, "hard keratins or β-keratins" and "soft keratins or α-keratins." In the hard tissue protein of nails, bird feathers, and fish scales that are comprised of short crystalline fibers enclosed in a strongly cross-linked elastomeric matrix they

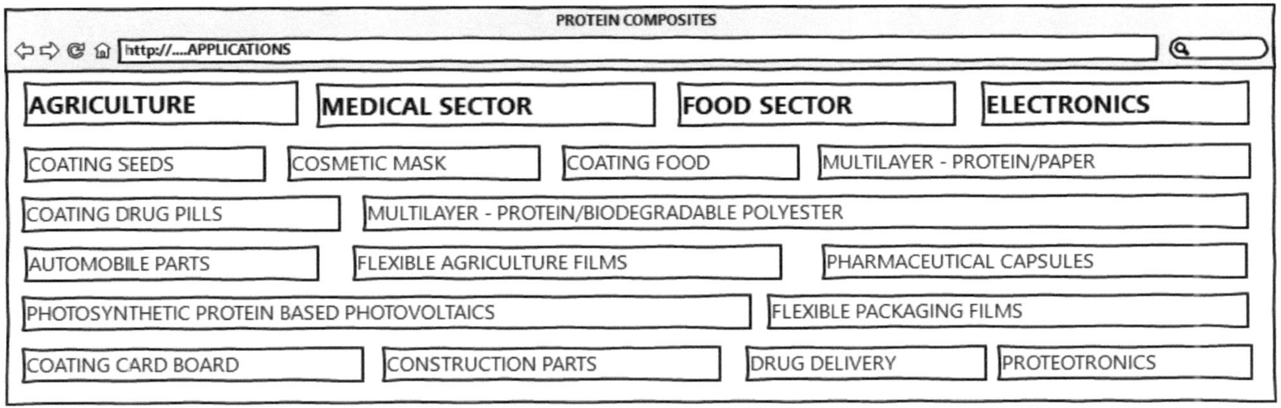

Figure 7.8 Applications of protein composites.

are rich in alanine and glycine, poor in cysteine, hydroxyproline, and proline. Cytoskeletal elements present in epithelial tissues such as the last layer of skin (epidermis), hair, and wool are soft keratins or α-keratins. These soft keratins are rich in cysteine that can be cross-linked through intermolecular disulfide bonds and offers strong and durable properties to hair [4, 5]. The structures of both forms of keratin are identical in the coiled-coil conformations, but their amino acid compositions are different [7]. For human keratins, molecular weights range from 44 to 66 kDa [8]. In polar and nonpolar solvents, keratins have high stiffness, strength, and insolubility. Intramolecular and intermolecular disulfide cross-links, hydrogen bonding, and their crystallinity are the result of stabilization. By these properties, it can distinguish from other fibrous proteins such as myofibrillar and collagen protein [9]. To utilize keratin as a green polymer, its extraction from biomass is required.

In the medical, pharmaceutical, cosmetic, and biotechnological industries, keratin is a very valuable product. Keratin-derived materials can be converted to the porous foam of various shapes, sponges, coatings, sheets, microfibers, gels, and high molecular weight materials. Keratin is gaining interest amongst researchers owing to its favorable properties such as biocompatibility, biodegradability as well as good mechanical properties. In designing wound healing gels, tissue engineering, drug delivery, films, sponges, trauma and medical equipment, biomedical, and cosmetic applications, keratin biomaterial is utilized. The production of biomaterials for regeneration and tissue repair is one of the immediate applications of purified keratin [10]. Balaji et al. fabricated three-dimensional composite scaffolds using a highly porous well interconnected matrix using keratin/gelatin and keratin/chitosan for tissue engineering and other biomedical applications. The scaffolds showed higher porosity with interconnected pores ranging from 20 to 100 μm. From the results, it was revealed that the prepared composites were a potential candidate for tissue engineering and other biomedical applications [11]. Tanabe et al. prepared a composite film using keratin as matrix and chitosan in 75% acetic acid. From the prepared green composites, it was demonstrated that chitosan could significantly alter the properties of the keratin film. These modifications particularly enhanced waterproof properties, which are considered to be beneficial for the application of the film generated from these two biopolymers to the medical sector [12]. Li et al. used keratin as filler in poly(L-lactic acid) (PLLA) matrix for nonwoven fibrous membrane by electrospinning the blend solutions. The results showed the addition of keratin increases the interactions between osteoblast cells and the polymeric membrane, as well as cell proliferation [13]. In another study, Li et al. produced a series of keratin and hydroxyapatite (HAP) nanocomposites by wet coprecipitation process at room temperature. X-ray diffraction (XRD) and transmission electron microscopy (TEM) verified that the crystallinity of HAP was reduced by the keratin matrix in the composite. The results of biocompatibility indicated that cells showed better viability on composites of keratin/HAP that have an organic and inorganic ratio close to that of natural bones [14]. Recently, a keratin and chitosan filler based 3D printed scaffold was successfully fabricated by Martinez et al. with polylactic acid (PLA) matrix. Effects of keratin fibers and particles were studied. Chitosan and keratin in the PLA matrix facilitate substantial increases in cell growth. The result concluded that the morphology of the reinforcement used changes the biocompatibility and thermomechanical properties [15]. In another recent study, keratin from chicken feather fiber and microcrystalline cellulose filler based PLA films were demonstrated by Khaw et al. They studied the effect of ionic liquid (IL) of 1-butyl-3-methylimidazolium chloride (BMIMCl) in the composite film. From scanning electron microscopy, it was observed that BMIMCI improved the miscibility of the composites. Thermal analysis revealed a decrease in the glass transition temperature (T_g) and crystallization temperature (T_c) with increase in the keratin composition [16]. Keratin reinforced poly(ε-caprolactone) (PCL) based composite material can be used in biomedical applications. Up to 30 wt% of keratin reinforced in PCL composites showed good mechanical properties and good uniformity in fiber morphology [17].

Zein accounts for about 80% of the protein content of corn. Zein is an alcohol-soluble protein (due to the presence of nonpolar amino acids in zein) that can be derived from agricultural derivatives such as corn gluten meal (CGM), dry milled corn (DMC), and distillers grains (DG).

In the medical, pharmaceutical, and food industries, zein has made recent advances in acting as a biomaterial. It has significant features, such as biodegradability, biocompatibility, mechanical efficiency, and excellent capacity for fiber and film forming ability. Zein is amphiphilic in nature, so it has properties that are hydrophobic and hydrophilic [18]. Zein is a combination of proteins that range in size and solubility of molecules. By different solubility and structures, zein proteins can be divided into four different types, such as α, β, γ, and δ. Considerably α-zein is the most generous; it is approximately 70% of the total protein and can be derived by using only aqueous alcohol [19].

Owing to its biodegradability, biocompatibility, and good support to multiple cell lines, it has gained significant consideration in biomedical applications [20]. For adhesives, binders, and pill capsules, zein protein is presently being used. Because of the properties such as good toughness, hydrophobic, glossiness, biodegradability, flexibility, and microbial resistant, zein-based composite films were gaining considerable attention in packaging applications [21]. Unplasticized zein films were brittle in nature, but the tensile strength improved two to three times after adding cross-linking agents such as formaldehyde, citric acid, and butanetetracarboxylic acid. By inserting highly stable silicate complexes into protein structures, zein films can be reinforced and modified [22]. Corn zein based films were prepared by Padgett et al., and the films showed improved strength and lower gas permeability than unmodified films. By integrating food-grade antimicrobial compounds into the packaging film, the marketability of zein-based plastic films can be enhanced [23]. Ghanbarzadeh and Oromiehi described the mechanical and film forming properties of single and laminated green composite films based on zein and whey protein with glycerol and olive oil used as a plasticizer. Whey protein films have strengthened mechanical and barrier properties by laminating them with zein-based films. Atomic force microscopy (AFM) found that zein films plasticized with olive oil had smoother surfaces as compared with other films prepared. The laminated films had a smoother surface than the single whey protein films, and despite that the laminated films were slightly rougher than single zein films. The laminated films showed higher ultimate tensile strength (UTS) than the single whey protein films, that is, 260% in whey/zein/glycerol and 200% in whey/zein/olive oil films. Compared to single whey protein films, the laminated films displayed greater barrier properties, that is, 180% in whey/zein/glycerol films and 200% in whey/zein/olive oil films and smaller than single zein films [24]. Altan et al. used zein as filler in PLA/zein fibrous composites film through electrospinning. Three different concentrations of carvacrol in the fibrous composites film were used. They concluded that the antioxidant activity of zein fibers loaded with carvacrol was from 62% to 75%, while the antioxidant potential of PLA fibers was from 53% to 65% with a content of 5–20% carvacrol. Sustained diffusion controlled release behavior was shown by the prepared fibrous composites film. The film also showed improved thermal stability. The effectiveness of carvacrol-loaded electrospun zein and PLA fibers on whole wheat (as a model food) showed antioxidant and antimicrobial properties, thereby indicating applicability in active food packaging [25]. Polycaprolactone (PCL) based composites scaffolds were fabricated by utilizing gum Arabic (GA) and zein as filler via three different electrospinning methods for skin regeneration applications. The prepared scaffold showed high porosity, hydrophilicity, and antibacterial activity. Good biocompatibility of the composites scaffold was revealed by in vitro cell culture [26]. Thermoplastic zein (TZ) was prepared by mixing zein with poly(ethylene glycol). Salerno et al. fabricated PCL-based biodegradable porous scaffold for bone tissue engineering through a supercritical CO_2 foaming process by adding HAP particles and TZ filler. The structural features of the prepared scaffold are affected by HAP concentration and foaming temperature [27].

Gluten is a group of proteins found in cereal grains and it is a renewable and biodegradable agricultural resource. Wheat gluten (WG) is a by-product of wheat starch and it comprises monomeric gliadin and polymeric glutenin. The viscosity of the gluten is influenced by gliadin, while the amount of glutenin will change elasticity.

The chemical composition of WG is heterogeneous and highly complicated. In this case, the terms monomeric and polymeric are related to the quaternary structure of the proteins. Gliadin is defined by the heterogeneous mixture of single chain or monomeric gluten protein with molecular weight (M_w) ranging between 30,000 to 74,000 and glutenin protein comprises more peptide chains connected with interchain disulfide bonds with M_w between 80000 to several million [28, 18]. WG has been gaining attention in different fields due to its thermoplastic and viscoelastic properties. WG is commonly used in both food and nonfood applications, which include cosmetics and hair products, detergents, rubber, and polymer products. These proteins possess good stability to water and heat, excellent elasticity, and good degradability. To produce biodegradable polymers from renewable resources, nonfood applications of wheat gluten have been investigated. WG was reinforced with unidirectional basalt fiber by Ye et al. to investigate the biodegradation behavior of the prepared green composites. Notably biodegradable behavior of the prepared composites was tested in soil (with prescribed moisture), whereby a steady decline in mechanical properties and mass was observed within 40 days [29]. El-Wakil fabricated green composites out of alkalized lignin and sodium silicate modified WG with and without silica gel and were characterized. It was confirmed that the addition of silica gel led to improved mechanical and thermal stability, higher glass transition temperature, low thermal expansion, higher tensile strength, and dimensional stability [30]. Kim introduced a novel technique for the development of green composite materials based on gluten at room temperature. This new strategy is economical as it reduces the use of chemicals and processing time; the end product is comparable to the commercial product, indicating its potential to partially substitute existing petroleum-based polymers [31]. Since WG has relatively low cost and good processability, it can be used as filler in aliphatic polyester-based composites. In order to reduce the cost of aliphatic polyester, WG is molded with aliphatic polyester. The WG-based aliphatic polyester composites showed good physical and thermal properties [32]. WG was added into PCL as a filler to produce biodegradable composites. Microscopic analysis showed the good dispersion of the filler in the matrix. The elongation at break was less than 100% at 50 wt% of filler and elongation was >900% until the 35 wt% of filler content. At a measured critical volume of 0.3 corresponding to 30% WG by weight with respect to PCL, there was a particle-induced transition [33]. Gluten composites were developed with unidirectional flax fibers. For stabilization, the composites were kept in a prepreg. Fiber length and the use of plasticizers are essential factors to enhance the mechanical properties. At 35% fiber volume fraction, a flexural strength of 117 MPa, the flexural modulus of 12 GPa, and 1.3% of strain to failure were received [34]. Owing to the film forming properties of wheat gluten, it can be used to produce plastic materials. Hemp fibers were added to the wheat gluten matrix to produce green plastics through compression molding. The mechanical properties increase due to the reinforcement of hemp fibers. A strong positive relation with tensile strength and Young's modulus and a negative relation with fracture strain and strain at full stress were shown by the fiber content in the material. Hemp fiber quality did not play an important role in tensile strength and strain, but Young's modulus correlated strongly and positively with the quality of hemp fiber. High performance liquid chromatography (HPLC), confocal laser scanning microscopy (CLSM), and SEM analysis suggested that relative to samples without diamine, the diamine added samples displayed a more "plastic" appearance along with a stiffer and stronger structure with less cracking [35]. Green composites were prepared from WG matrix and renewable resources as reinforcement to solve the management problem related to algae accumulation by the hot press molding method. From the research, it was concluded that hot press molding is an efficient method to develop sustainable composites derived from algae waste [36].

Peanut is an annually grown plant and it is the third most essential source of plant protein. Worldwide, 11% of protein supply comes from peanuts and it is grown on approximately 42 million acres. It is a cost-effective and essential source of plant protein. Almost 70% of proteins of good nutritional quality are found in peanuts and most of the proteins are in the storage form [37, 38].

There are primarily three types of proteins present in peanuts, namely, arachin, conarachin I, and conarachin II, which make up to 75% of the peanut total proteins. In different fields, arachin and conarachin find a wide range of uses; the most important is in the food industry [39]. The arachin and conarachin both are proteins and their amino acid chain folding in a way that improves the solubility of the protein in water by putting polar groups of atoms at the surface of the protein allowing nonpolar side interaction at the center, thereby creating globular domains, and they are specifically glycoprotein having neutral amino sugar moiety, where the sugar content is very less [40]. In their structure, both arachin and conarachin have a significant portion of tryptophan, tyrosine, and phenylalanine fluorescent amino acids [39]. As it comes from a plant source, the properties of peanut proteins are different from animal proteins, which include fiber and bioactives such as arginine and resveratrol. Resveratrol can help to decrease cardiovascular disease, cancer, and inflammation and arginine can help to relax our arteries, lower blood pressure, and improve blood flow by converting blood sugar into energy and turning it into nitric oxide [37]. Different solvents such as sodium chloride, deionized water, and calcium chloride were used to extract peanut protein from defatted peanut flour [41].

Peanut proteins can be utilized in the production of films owing to their superior physical and mechanical properties. Zhong et al. prepared peanut protein isolate (PPI) based on active films by incorporating thymol (TML) into it. Glycerol was added as a plasticizer in the PPI matrix. PPI has better functional properties than other peanut products and has higher protein content. The addition of TML increases the antioxidant property of the films. It was concluded that the addition of a required amount of TML in PPI films can preserve food and can be utilized for active food packaging applications [42]. Sun and Xiong investigated the properties of protein starch [PS] and PPI blend films. The addition of PPI into the PS matrix at 50 wt% increased the elongation from 28.56% to 98.12% but decreased the tensile strength from 5.44 to 3.06 MPa. The blend upon addition of 20 and 50 wt% was not homogeneous. At PPI 40 wt%, the smoother surface was seen in PS/PPI blend films [43]. Peanut protein cross-linked with citric acid showed composites films with good dry and wet tensile properties [44].

Fish proteins are derived from marine resources, including fish processing by-products. These are excellent sources of high-quality protein and contain essential amino acids. Fish proteins typically contain all the essential amino acids, especially high content of lysine and leucine. Several amino acids found in fish proteins are known to modify different metabolic pathways, that is, reducing hypertension by inhibiting the angiotensin-1 converting enzyme. Fish proteins can thus contribute to disease prevention [45]. Green composites with desirable properties can be derived by using fish proteins for different applications. Recently Lin et al. derived fish protein based composites film for food packaging applications. They used fish gelatin (FG) and chitosan as a film forming substrate and silver-loaded nano-titania (TiO_2Ag) was added to the composites films to improve the antibacterial properties. The smooth and uniform surface of the prepared composites was confirmed by field emission scanning electron microscopic (FE-SEM) analysis. From the structural analysis, it was concluded that the presence of TiO_2Ag crystals in the film could improve the interaction between the components [46]. Fish gelatin can be utilized to produce biodegradable films. Fish gelatin films were prepared by the addition of buriti oil and glycerol as a plasticizer. The films produced had high antioxidant activity (24.77 μmol TE/g). The fish gelatin based film showed a mechanical resistance of 8.04 MPa and water vapor permeability (WVP) of 0.38×10^{-10} g. m. m^{-2}. s^{-1}. Pa^{-1} [47]. Owing to the triple helix structure of the collagen derived from fish, it is able to improve the mechanical and thermal properties of PLA compared with other polymers. Fish collagen was used as a filler to improve the mechanical and thermal properties of PLA through a condensation reaction. After the grafting of collagen into the PLA matrix, the tensile strength increased by 88.60% and the elongation at break improved by 176.88%. When compared with pure PLA, the enthalpy increases in the case of PLA/collagen materials. From the research, it was concluded that fish proteins had a great plasticizing effect and can be useful to develop high impact resistance of PLA-based composites to expand the application fields of PLA materials [48]. Im et al. used fish collagen as a coating agent to prepare a PCL/fish collagen/alginate/phlorotannin-based composite scaffold for tissue engineering applications. First, they printed a PCL scaffold through additive manufacturing and then phlorotannin and fish collagen/alginate were applied by a simple coating method. The coating agents on the scaffolds provided a good hydrophilic surface and induced meaningful wetting ability [49]. In another study, fish collagen along with nano-hydroxyapatite (N-HAP) was used as filler to improve the mechanical properties, desired degradation rate, cytocompatibility, and osteogenic activity of poly(lactide-*co*-glycolide) (PLGA) based scaffold through electrospinning. The morphological and physicochemical properties revealed that fish collagen and N-HAP were uniformly distributed in the matrix. Because of the interaction between fish collagen and PLGA, the tensile strength of the composite was significantly improved [50]. In another research, Hosseini et al. used fish gelatin as filler in PLA-based green composites multilayer films via the solution casting technique. The SEM results revealed that the PLA layers are closely attached to the inner fish gelatin layers to make a continuous film. The tensile strength of the triple layer film was higher than that of fish gelatin. The water vapor permeability and oxygen permeability of the prepared multilayer film were reduced to 91% and 87%, respectively. This multilayer film is environmentally friendly and has the potential to substitute the synthetic oxygen barrier components in multilayer packaging formulations like EVOH and hence has a high versatility in food packaging applications [51].

Silks proteins are extremely strong and tough fibrous proteins that are generated in the form of fibers from silkworms, spider scorpions, and some insects. Within specialized glands, silk proteins are widely produced by various species, including silkworms, scorpions, spiders, mites, and flies.

Silk consists of two separate elements, that is, the core protein fibroin and its adhesive-like coating sericin that is extracted before the fiber is converted into useful materials. In epithelial cells, these proteins are biosynthesized and secreted into the lumen of these glands, where the proteins are processed before being woven into silk fibers. Based on their particular natural functions, various silk proteins vary in terms of structure and properties, amino acid composition and different silks are synthesized for different functions in various spiders [52, 53]. For protein composition and amino acid sequence, *Bombyx mori* and dragline silk from the spider N. clavipes and cocoon silk fibroin from the silkworm have been the most widely analyzed. Sericins are more hydrophilic than fibroin, which is why they are quickly separated from cocoons by boiling silk for textile applications in alkaline solutions. This action helps the fibroin to be spooled in. Other silks have also been established, like caddis fly and aquatic midge, which spin silks underwater to form sheltered tubes [54]. Based on the thermal gravimetric analysis (TGA), spider dragline silk is thermally stable to about 230°C. A dynamic mechanical study observed two thermal transitions, one at –75 °C, assumed to reflect localized mobility in the noncrystalline regions of silk fiber, and another at 210°C, assumed to be representative of partial melt, glass transition (T_g), or amorphous to β transitions. Silkworm silk has thermal properties equivalent to those of spider silk [55]. Silk proteins are considered and rediscovered as one of the most fascinating and promising naturally originated materials for different biomedical and other technological applications because of their extraordinary mechanical properties, environmental stability, biocompatibility, and controlled degradability among other properties. Combined with other natural/synthetic polymers or even inorganic phases, various protein-based composite materials (films, fibers, foams, gels, etc.) can be made. Hu et al. prepared silk fibroin filler based green hybrid composite by combining silks with poly(L-lactide) (PLLA) biodegradable polymer with the purpose of biodegradable disposal. The hybrid composite scaffold was prepared by freezing and lyophilizing a blend of fibroin microspheres and PLLA solution. Fourier transform infrared spectroscopy (FTIR) and scanning electron microscopy (SEM) analysis revealed that presence of fibroin microspheres was on the surface of the hybrid scaffold. Compared to the PLLA scaffold, SEM and laser scanning confocal microscope (LSCM) analysis demonstrated that the human hepatocellular carcinoma HepG2 cells had proliferated and spread much more in the hybrid scaffold. Fibroins with PLLA showed increased tensile strength and elongation at break although hydrophilicity decreased [56]. A promising area is the synthesis of silk with other natural fibers. Hu et al. prepared a novel biocompatible film as a scaffold material by solution blending of recombinant human-like collagen (RHLC) and *B. mori* silk fibroin matrix for hepatic tissue engineering applications. Solution blending was utilized to combine RHLC with silk fibroin to improve the blend films' biocompatibility and hydrophilicity while retaining and even improving flexibility. From contact angle measurement, the surface of the RHLC/fibroin film was more hydrophilic. FTIR and XRD analysis revealed that hydrogen bonds had formed between fibroin and RHLC and the SEM analysis data confirmed that homogeneous microstructures were still retained after the introduction of RHLC with fibroin. While the number of H-bonds increased because of the presence of collagen in the films, when compared to pure fibroin the mechanical properties were lower [57]. Recently, *B. mori* silk microparticles (SMP) filler based PCL composites scaffolds were prepared through extrusion-based 3D printing technique. The addition of SMP in the PCL matrix significantly improved the compressive Young's modulus and rate of degradation of the composite scaffolds [58]. Electron beam treated silk fiber reinforced poly(butylene succinate) (PBS) composites were demonstrated for the first time. It was concluded that the use of electron beam irradiation at an acceptable dosage may contribute to improving the thermodimensional stability and dynamic mechanical and flexural properties of the green composites based on silk [59]. Silk can be used in internal bone clips for fractures. PLA, HAP, and silk based in the bone clip for internal fixation of bone fracture were developed by 3D printing. Different properties of 3D printed PLA, PLA/HAP, and PLA/HAP/silk composites bone clips were compared and characterized. PLA/HAP/silk composites bone clip showed the same mechanical properties. The addition of silk along with HAP improves the biocompatibility of PLA/HAP/silk composites [60].

Collagen is the most generous structural protein found in mammalian tissues and it is an important part of the extracellular matrix. It is found throughout the body. It is the important component of skin, tendon, bone, ligament, cartilage, and most internal organs by supplying structural and mechanical support to the softer tissues and connecting them to the skeleton.

With a very small molecular weight distribution (approximately 200 and 100 kDa for β and α chains, respectively) [61], this fibrous and structural protein includes 28 distinct forms of collagen consisting of more than 40 separate polypeptide chains. By non-covalent interactions, the individual chains are retained in a triple helical configuration. Three polyproline polypeptide chains consist of the triple helical structure of collagen [62, 63]. These polypeptides have repeated X-Y-Gly sequences in which proline (Pro, P) and its posttranslational modification 4-hydroxyproline (Hyp, O) are frequently occupied by X and Y positions, respectively. Of the 28 different forms of collagen, the primary ones are fibril forming collagen, that is, type I (e.g., bone, skin, tendon), type II (e.g., cartilage), and type III (e.g., blood vessels, skin). Generally, these collagen varieties belong to fibrillar structures that create an essential part of the architecture and integrity of the tissues. It is suspected that the presence of hydroxyproline in collagen leads to its thermostability; mutations that eliminate hydroxyproline from different points in the collagen sequence greatly reduce the temperature of its thermal denaturation [64]. Whole body protein consists up to 25–35% of the various collagens [65].

Collagen is also found in mineralized tissues like teeth or bones, where it interacts with the much more complicated phases of hydroxyapatite [66]. Therefore, the function of collagen here is to give flexibility, assuring higher resistance to fracture. Collagen matrices, in addition to these tissues, play an essential role in cartilage, skin, blood vessels, and muscle flexibility. Collagen has Young's modulus of about 1.2 GPa, toughness of 6 MJ/m^3, and strength of 120 MPa in terms of mechanical properties. Since their extensibility is usually 13–15%, collagen fibers are also described as inelastic. Although collagen is able for reversible deformation, that also contributes to its description as elastin protein [67].

In the biomedical area, collagen has been broadly researched, demonstrating interesting properties for the production of biocomposites, biofilms, and tissue engineering scaffolds, particularly in reconstructive medicine, pharmaceuticals, and cosmetics. For cell attachment, proliferation, and differentiation, collagen can act as a natural substrate and makes it a suitable matrix material for applications in tissue engineering and wound dressing [63]. Kumar et al. prepared collagen/chitosan scaffolds of various compositions through an emulsion air drying method. The addition of 10–30 wt% of chitosan as filler in collagen matrix resulted in better mechanical properties having a tensile strength of 13.57 MPa with 9% elongation at break. This research showed that a composite scaffold can be produced using the emulsion air drying process with two entirely different natural polymer systems. The collagen and chitosan-made composite scaffolds are biocompatible and have sufficient physical and structural characteristics to facilitate the attachment and proliferation of fibroblast, indicating that it is a successful substrate for the biomedical application [68].

Negishi et al. used collagen as a modifier PLA-based scaffold by the vacuum pressure impregnation (VPI) method. First, the PLA scaffold was prepared and then collagen solution was introduced into the PLA scaffold as filler by using the VPI method, which was not possible using an immersion treatment. When compared with the neat PLA scaffold, the prepared PLA/collagen composite scaffold had greater water adsorption and degradation. These findings suggest that VPI could be a promising method for the production of natural material composites [69]. Yeo et al. fabricated collagen protein filler based three-dimensional hierarchical composites scaffold of PCL/β-tricalcium phosphate (β-TCP) and collagen nanofibers to determine the physical properties and in vitro cell activity for bone tissue regeneration. From the morphological analysis, it was concluded that the electrospun collagen nanofibers are well layered between the composites struts and β-tricalcium phosphate was uniformly dispersed in PCL struts. The addition of collagen nanofibers and β-TCP increases the water absorption ability and increases the hydrophilicity of the three-dimensional hierarchical composites scaffold [70]. In another study, Wei et al. prepared a gelatin/collagen/PCL (G/C/P) based scaffold for skin regeneration. G/C/P biocomposites were prepared by impregnating lyophilized G/C with PCL solutions. Two different ratios of G/C:PCL were used in this study. The thermal analysis showed that the G/C/P scaffold had characteristic melting point of PCL around 60°C. All biocomposites showed the same fibrous structure. The biocomposites with lower content of collagen and higher content of PCL were chosen for wound healing studies because of their reasonable mechanical strength [71]. Collagen can also be used in cartilage regeneration. PCL/collagen-based scaffolds were successfully produced by using PCL and different concentrations of collagen as a filler by thermally induced phase separation method with advantages such as bioactivity of collagen and mechanical properties of PCL [72]. PCL, nano-hydroxyapatite (n-HAP), and collagen hybrid scaffolds were fabricated by a combination of solvent casting, particulate leaching, and polymer leaching process. PCL and n-HAP scaffolds were synthesized and chemically modified. Collagen was used as filler and it was grafted onto the surface of the PCL/n-HAP scaffold. The study highlights the value of collagen grafting on PCL/n-HAP scaffolds in modulating cellular activity and biological functions, extending its current application from soft tissue engineering to hard tissue engineering [73]. PLA as matrix and collagen as filler-based blends were produced by Yang et al. with the improved biodegradable property. From the analysis, it was concluded that with increasing the collagen content, the glass transition temperature (T_g) and crystallinity of the blends decreased and the enzymatic hydrolysis rate of PLA increased [74]. Zhijiang and Guang prepared bacterial cellulose (BC)/collagen composite scaffolds with a foam structure by immersing BC films in a collagen solution followed by freeze-drying. The surface morphology revealed that the collagen not only coats the BC membranes but also penetrates into them as filler. The prepared scaffolds showed enhanced mechanical properties and lower elongation and improved cytocompatibility when compared with pure BC scaffolds [75]. Mathew et al. investigated the collagen and cellulose nanofibers based implantable composites by solution casting followed by pH-induced in situ partial fibrillation of collagen and cross-linking using glutaraldehyde. From the microscopy study, it was revealed that fibrous collagen and cellulose nanofibrils were inserted in the collagen matrix. The addition of cellulose nanofibrils and cross-linking showed a considerable effect on the mechanical properties of the nanocomposites, that is, tensile strength 132–150 MPa and modulus 5–6 MPa. In order to adapt the mechanical properties and cytocompatibility of these composites for particular applications like artificial ligaments or tendons, the collagen/cellulose ratio, cross-linking agent, and cross-linking level of collagen may be optimized further [76].

In bovine milk, casein is the primary protein present at 24–29 g/L, and it is acquired by precipitating milk at pH 4.6, with the supernatant fraction being whey protein, another major component in the milk protein. Casein is also considered an inherently unstructured protein without any particular secondary structure or tertiary structure with open and flexible conformation. Caseins are often random coil polypeptides with a high degree of molecular flexibility that can form typical intermolecular interactions due to the low frequency of secondary structures (α-helix and β-sheets). Depending upon the coagulation process, two varieties of caseins are available: acid and rennet caseins. The four major components of casein are α_{s1}-casein, α_{s2}-casein, β-casein, and κ-casein that make up to 38%, 10%, 36%, and 13%, respectively, of casein composition. Each component has a different molecular weight, amino acid composition, isoelectric point, and hydrophilicity. In recent years, casein has been widely investigated for different applications due to its physical and structural properties, and the physical and structural properties of casein make it a material with superior functionality [77, 78]. Caseins are insoluble when compared with other proteins that make caseinate (mostly sodium caseinate), and this is also utilized as a substitute for conventional packaging material [79]. Milk casein was used as a dispersant in PLA and cellulose nanowhiskers (CNW) composites due to its ability to interact with PLA and CNW. Before making the composites, CNW fillers were modified using casein. Casein enhanced the dispersion of CNW in the PLA matrix [80]. In order to improve the fire resistance and biodegradability, Zhang et al. incorporated casein in the PLA matrix by melt compounding process. Casein increased the fire retardancy of the PLA composites. With the addition of 20% of casein, the limiting oxygen index (LOI) values of PLA composites increased from 20.0% to 32.2%, upgraded the UL-94 rating from no rating to V-0, and the peak heat release rate reduced from 779 to 639 kW/m^2 [81]. Caprolactam (CPL) modified casein-based composites films were successfully fabricated with the addition of a waterborne polyurethane (WPU) film forming binder and characterized. Good dispersion of WPU in the casein matrix was observed on the addition of 50% of WPU, but the compatibility between casein-CPL and WPU became poorer while increasing the WPU amount above 50% [82]. Casein-based nanocomposites can also be used as antibacterial coating with a great potential application in various fields such as packaging, textile, paper making, leather, and indoor wall coating. Casein-based flexible hybrid composites films were synthesized by the in situ polymerization process. In order to improve the flexibility, extensibility, and antimicrobial property of the casein film, CPL and zinc oxide nanoparticles (ZnO NPs) were added to the casein matrix. From the analysis, it was concluded that casein-based films showed good antibacterial activities against *Staphylococcus aureus* and *Escherichia coli*. The mechanical strength of the casein film increased only with 0.5 wt% of ZnO NPs [83].

7.6 Conclusion

Protein-based composites offer countless possibilities to customize material properties by manipulating the protein sequences from which they are produced or by which they are manufactured. The range of possible applications is addressed in this chapter. In comparison to other natural polymers such as lignocellulosic components (cellulose, hemicellulose, and lignin), starch with respect to intrinsic biodegradability, biocompatibility, natural abundance, and mechanical and functional properties, proteins have shown promising results to provide materials for next generation. To conclude, the utilization of eco-friendly materials with enhanced characteristics gives a useful opportunity to partially substitute conventional plastics from petroleum resources, giving more value to the vast naturally available macromolecules.

Acknowledgments

The second author is thankful to CSIR and DST – South Africa for providing the CSIR – DST Inter-Programme Bursary Award in the area of Composites for the years 2016, 2017, and 2018. The authors also extend thanks to Durban University of Technology, South Africa, and Mahatma Gandhi University, Kerala, India for their support to carry out this work. The authors also wish to thank all the authors whose work has been referred to while writing this chapter.

References

1. C. Bastioli, *Handbook of Biodegradable Polymers*, Rapra Technology Limited, Shawbury, 2005.
2. K. Matyjaszewski, Protein Folding, in: *Encyclopedia of Polymer Science and Technology*, 4th ed., John Wiley & Sons, 2002: pp. 644–656. doi:10.1002/0471440264.
3. K. McGrath, D. Kaplan, Protein-Based Materials, Birkhauser, Boston, 1997. doi:10.1007/978-1-4612-4094-5.
4. J.M.C. Kittrick, P. Chen, S.G. Bodde, W. Yang, E.E. Novitskaya, M.A. Meyers, The structure, functions, and mechanical properties of keratin, *J. Mater.* 64 (2012) 449–468. doi:10.1007/s11837-012-0302-8.
5. J.G. Rouse, M.E. Van Dyke, A review of keratin-based biomaterials for biomedical applications, *Materials (Basel)* (2010) 999–1014. doi:10.3390/ma3020999.
6. P.A. Coulombe, M.B. Omary, 'Hard' and 'soft' principles defining the structure, function and regulation of keratin intermediate filaments, *Curr. Opin. Cell Biol.* 14 (2002) 110–122.
7. H. Lee, K. Noh, S.C. Lee, I. Kwon, D. Han, I. Lee, Y. Hwang, Human hair keratin and its-based biomaterials for biomedical applications, *Tissue Eng. Regen. Med.* 11 (2014) 255–265. doi:10.1007/s13770-014-0029-4.
8. R. Moll, M. Divo, L. Langbein, The human keratins: Biology and pathology, *Histochem. Cell Biol.* 129 (2008) 705–733. doi:10.1007/s00418-008-0435-6.
9. P.M.M. Schrooyen, P.J. Dijkstra, R.C. Oberthu, A. Bantjes, J. Feijen, Partially carboxymethylated feather keratins. 1. Properties in aqueous systems, *J. Agric. Food Chem.* 48 (2000) 4326–4334.
10. P. Kumaran, A. Gupta, S. Sharma, Synthesis of wound-healing keratin hydrogels using chicken feathers proteins and its properties, *Int. J. Pharm. Pharm. Sci.* 9 (2017).

11. S. Balaji, R. Kumar, R. Sripriya, P. Kakkar, D.V. Ramesh, P.N. Kanta, P.K. Sehgal, Preparation and comparative characterization of keratin – chitosan and keratin – gelatin composite scaffolds for tissue engineering applications, *Mater. Sci. Eng.* C. 32 (2012) 975–982. doi:10.1016/j.msec.2012.02.023.
12. T. Tanabe, N. Okitsu, A. Tachibana, K. Yamauchi, Preparation and characterization of keratin – chitosan composite film, *Biomaterials.* 23 (2002) 817–825.
13. J. Li, Y. Li, L. Li, A.F.T. Mak, F. Ko, L. Qin, Preparation and biodegradation of electrospun PLLA/keratin nonwoven fibrous membrane, *Polym. Degrad. Stab.* 94 (2009) 1800–1807. doi:10.1016/j.polymdegradstab.2009.06.004.
14. J. Li, X. Liu, J. Zhang, Y. Zhang, Y. Han, J. Hu, Y. Li, Synthesis and characterization of wool keratin/hydroxyapatite nanocomposite, *J. Biomed. Mater. Res. Part B.* 100B (2012) 896–902. doi:10.1002/jbm.b.32645.
15. L.E. Rojas-Martínez, C.G. Flores-Hernandez, L.M. López-Marín, A.L. Martinez-Hernandez, 3D printing of PLA composites scaffolds reinforced with keratin and chitosan: Effect of geometry and structure, *Eur. Polym. J.* 141 (2020) 110088. doi:10.1016/j.eurpolymj.2020.110088.
16. Y.Y. Khaw, Y.C. Ching, N. Liu, S.N. Gan, R. Singh, N. Nazri, N. Ghazali, Poly (lactic acid) composite films reinforced with microcrystalline cellulose and keratin from chicken feather fiber in 1-butyl-3-methylimidazolium chloride, *J. Appl. Polym. Sci.* 47642 (2019) 1–10. doi:10.1002/app.47642.
17. A. Edwards, D. Jarvis, T. Hopkins, S. Pixley, N. Bhattarai, Poly(e-caprolactone)/keratin-based composite nanofibers for biomedical applications, *J. Biomed. Mater. Res. Part B.* 103B (2014) 21–30. doi:10.1002/jbm.b.33172.
18. E.J. Bealer, S. Onissema-karimu, A. Rivera-galletti, M. Francis, J. Wilkowski, X. Hu, Protein – Polysaccharide composite materials, *Polymers (Basel)* 12 (2020) 1–28.
19. H. Xu, Y. Yang, 3D Electrospun fibrous structures from biopolymers, in: *Lightweight Materials from Biopolymers and Biofibers*, American Chemical Society, Washington, 2014: pp. 103–126.
20. Q. Jiang, Y. Yang, Water-stable electrospun zein fibers for potential drug delivery, *J. Biomater. Sci. Polym. Ed.* 22 (2011) 1393–1408. doi:10.1163/092050610X508437.

21. R. Shukla, M. Cheryan, Zein: The industrial protein from corn, *Ind. Crops Prod.* 13 (2001) 171–192.

22. H.J. Park, et al., Silicate modification of corn protein films, *J. Mater. Sci. Lett.* 7 (1998) 359–361.

23. T. Padgett, I.Y. Han, P.I. Dawson, Incorporation of food-grade antimicrobial compounds into biodegradable packaging films, *J. Food Prot.* 61 (1998) 1330–1335.

24. B. Ghanbarzadeh, A.R. Oromiehi, Biodegradable biocomposite films based on whey protein and zein: Barrier, mechanical properties and AFM analysis, *Int. J. Biol. Macromol.* 43 (2008) 209–215. doi:10.1016/j.ijbiomac.2008.05.006.

25. A. Altan, Z. Aytac, T. Uyar, Carvacrol loaded electrospun fibrous films from zein and poly(lactic acid) for active food packaging, *Food Hydrocoll.* (2018). doi:10.1016/j.foodhyd.2018.02.028.

26. Z.P. Rad, J. Mokhtari, M. Abbasi, Calendula officinalis extract/PCL/Zein/Gum arabic nano fibrous bio-composite scaffolds via suspension, two-nozzle and multilayer electrospinning for skin tissue engineering, *Int. J. Biol. Macromol.* 135 (2019) 530–543. doi:10.1016/j.ijbiomac.2019.05.204.

27. A. Salerno, S. Zeppetelli, E. Di, S. Iannace, P.A. Netti, Novel 3D porous multi-phase composite scaffolds based on PCL, thermoplastic zein and ha prepared via supercritical CO_2 foaming for bone regeneration, *Compos. Sci. Technol.* 70 (2010) 1838–1846. doi:10.1016/j.compscitech.2010.06.014.

28. W.S. Veraverbeke, J.A. Delcour, Wheat protein composition and properties of wheat glutenin in relation to breadmaking functionality, *Crit. Rev. Food Sci. Nutr.* 42 (2010) 37–41.

29. P. Ye, L. Reitz, C. Horan, R. Parnas, Manufacture and biodegradation of wheat gluten/basalt composite material, *J. Polym. Environ.* 14 (2006) 1–7. doi:10.1007/s10924-005-8701-3.

30. N.A. El-wakil, Use of lignin strengthened with modified wheat gluten in biodegradable composites, *J. Appl. Polym. Sci.* 113 (2009) 793–801. doi:10.1002/app.

31. S. Kim, Processing and properties of gluten/zein composite, *Bioresour. Technol.* 99 (2008) 2032–2036. doi:10.1016/j.biortech.2007.02.050.

32. S.L. Wook, I.J. Kwon, K. Hee, B. Suk, Structure and properties of biodegradable gluten/aliphatic polyester blends, *Eur. Polym. J.* 35 (1999) 1875–1881.

33. V.L. Finkenstadt, A.A. Mohamed, G. Biresaw, J.L. Willett, Mechanical properties of green composites with polycaprolactone and wheat gluten, *J. Appl. Polym. Sci.* 110 (2008) 2218–2226. doi:10.1002/app.

34. S. Hemsri, A.D. Asandei, K. Grieco, R.S. Parnas, Biopolymer composites of wheat gluten with silica and alumina, *Compos. Part A* 42 (2011) 1764–1773. doi:10.1016/j.compositesa.2011.07.032.

35. C. Wretfors, S.-W. Cho, M. S. Hedenqvist, S. Marttila, S. Nimmermark, E. Johansson, Use of industrial hemp fibers to reinforce wheat gluten plastics, *J. Polym. Environ.* 17 (2009) 259–266. doi:10.1007/s10924-009-0147-6.

36. B. Ferrero, T. Boronat, R. Moriana, O. Fenollar, R. Balart, U. Polite, Green composites based on wheat gluten matrix and posidonia oceanica waste fibers as reinforcements, *Polym. Compos.* 34 (2013) 1663–1669. doi:10.1002/pc.

37. S.K. Ghatak, K. Sen, Peanut proteins: Applications, ailments and possible remediation, *J. Ind. Eng. Chem.* 19 (2013) 369–374. doi:10.1016/j.jiec.2012.09.009.

38. B. L Jani, B.M. Devani, Peanut protein: Rich source as vegan protein, *HSOA J. Food Sci. Nutr.* 6 (2020) 1–5. doi:10.24966/fsn-1076/100059.

39. P.V. Monteiro, V. Prakash, Effect of proteases on arachin, conarachin I, and conarachin II from peanut (*Arachis hypogaea* L.), *J. Agric. Food Chem.* 42 (1994) 268–273. doi:10.1021/jf00038a008.

40. A.K. Jain, Cloning and structural analysis of a cDNA clone encoding glycinin (Gly-1) seed storage protein of peanut, *Electron. J. Biotechnol.* 7 (2004) 221–231. doi:10.2225/vol7-issue3-fulltext-13.

41. T.S. Kholief, Chemical composition and protein properties of peanuts, *Z. Ernahrungswiss.* 26 (1987) 56–61. doi:10.1007/BF02023820.

42. T. Zhong, Y. Liang, S. Jiang, L. Yang, Y. Shi, S. Guo, C. Zhang, Physical, antioxidant and antimicrobial properties of modified peanut protein isolate based films incorporating thymol, *RSC Adv.* 7 (2017) 41610–41618. doi:10.1039/c7ra07444a.

43. Q. Sun, C. Sun, L. Xiong, Mechanical, barrier and morphological properties of pea starch and peanut protein isolate blend films, *Carbohydr. Polym.* 98 (2013) 630–637. doi:10.1016/j.carbpol.2013.06.040.

44. N. Reddy, Q. Jiang, Y. Yang, Preparation and properties of peanut protein films crosslinked with citric acid, *Ind. Crops Prod.* 39 (2012) 26–30. doi:10.1016/j.indcrop.2012.02.004.

45. H.F. Dale, L. Madsen, Fish-derived proteins and their potential to improve human health, *Nutr. Rev.* (2019) 1–12. doi:10.1093/nutrit/nuz016.

46. D. Lin, Y. Yang, J. Wang, W. Yan, Z. Wu, H. Chen, Q. Zhang, D. Wu, W. Qin, Z. Tu, Preparation and characterization of TiO_2-Ag loaded fish gelatin-chitosan antibacterial composite film for food packaging, *Int. J. Biol. Macromol.* 154 (2020) 123–133. doi:10.1016/j.ijbiomac.2020.03.070.
47. N. Da, S.E. Silva, E. José, G. Pino, C. Da, S. Araújo, M. Regina, S. Peixoto, L.D.F. Henriques, N. Da, S.E. Silva, E. José, G. Pino, D.S. Araújo, M. Regina, S. Peixoto, L. De Fátima, H. Lourenço, Development and optimization of biodegradable fish gelatin composite film added with buriti oil, *CyTA - J. Food.* 16 (2018) 340–349. doi:10.1080/19476337.2017.1406005.
48. H. Ma, J. Shen, Q. Yang, J. Zhou, S. Xia, J. Cao, Effect of the introduction of fish collagen on the thermal and mechanical properties of poly (lactic acid), *Ind. Eng. Chem. Res.* (2015). doi:10.1021/acs.iecr.5b02969.
49. J. Im, H. Choi, F. Mun, J. Lee, H. Kim, W.K. Jung, C.H. Jang, G. Kim, Biocomposite supplemented with phlorotannin for hard tissue regeneration, *RSC Adv.* (2017) 2009–2018. doi:10.1039/c6ra25182j.
50. S. Jin, F. Sun, Q. Zou, J. Huang, Y. Zuo, Y. Li, S. Wang, L. Cheng, Y. Man, F. Yang, J. Li, Fish collagen and hydroxyapatite reinforced poly (lactide-co-glycolide) fibrous membrane for guided bone regeneration, *Biomacromolecules* 20 (2019) 2058–2067. doi:10.1021/acs.biomac.9b00267.
51. S.F. Hosseini, Z. Javidi, M. Rezaei, Efficient gas barrier properties of multi-layer films based on poly (lactic acid) and fish gelatin, *Int. J. Biol. Macromol.* 92 (2016) 1205–1214. doi:10.1016/j.ijbiomac.2016.08.034.
52. G.H. Altman, F. Diaz, C. Jakuba, T. Calabro, R.L. Horan, J. Chen, H. Lu, J. Richmond, D.L. Kaplan, Silk-based biomaterials, *Biomaterials* 24 (2003) 401–416.
53. N.C. Abascal, L. Regan, The past, present and future of protein-based materials, *Open Biol.* 8 (2018) 1–8.
54. S.T. Case, J. Powers, R. Hamilton, M.J. Burton, Silk and silk proteins from two aquatic insects, in: D. Kalpan, W. Adams, B. Farmer, C. Viney (Eds.), *Silk Polymers: Material Science and Biotechnology,* American Chemical Society, Washington, 1994: pp. 80–90.
55. P.M. Cunniff, S.A. Fossey, M.A. Auerbach, J.W. Song, Mechanical properties of major ampulate gland silk Fibers extracted from Nephila clavipes spiders, in: D. Kalpan, W.W. Adams, B. Farmer, C. Viney (Eds.), *Silk Polymers: Material Science and Biotechnology,* Washington, DC, 1994: pp. 234–251.
56. K. Hu, Q. Lv, F.Z. Cui, L. Xu, Y.P.J. Wang, Q.L. Feng, H.L. Wang, L.Y. Huang, A novel poly (L-lactide) (PLLA)/fibroin hybrid scaffold to promote hepatocyte viability and decrease macrophage responses, *J. Bioact. Compat. Polym.* 22 (2007) 395–410. doi:10.1177/0883911507079893.

57. K. Hu, Q. Lv, F.Z. Cui, Q.L. Feng, X.D. Kong, H.L. Wang, L.Y. Huang, T. Li, Compatible polymers films with recombinant human-like collagen for hepatic tissue engineering, *J. Bioact. Compat. Polym.* 21 (2006) 23–37. doi:10.1177/0883911506060455.

58. C. Vyas, J. Zhang, Ø. Øvrebø, B. Huang, I. Roberts, M. Setty, 3D printing of silk microparticle reinforced polycaprolactone scaffolds for tissue engineering applications, *Mater. Sci. Eng. C.* 118 (2021) 111433. doi:10.1016/j.msec.2020.111433.

59. Y. Kim, O.H. Kwon, W.H. Park, D. Cho, Thermomechanical and flexural properties of chopped silk fiber-reinforced poly (butylene succinate) green composites: Effect of electron beam treatment of worm silk, *Adv. Compos. Mater.* 22 (2013) 37–41. doi:10.1080/09243046.2013.843830.

60. Y.K. Yeon, H.S. Park, J.M. Lee, J.S. Lee, Y.J. Lee, T. Sultan, Y. Bin Seo, O.J. Lee, S.H. Kim, C.H. Park, New concept of 3D printed bone clip (polylactic acid/hydroxyapatite/silk composite) for internal fixation of bone fractures, *J. Biomater. Sci. Polym. Ed.* 5063 (2017) 0–1. doi:10.1080/09205063.2017.1384199.

61. Z. Zhang, G. Li, B.I. Shi, Physicochemical properties of collagen, gelatin and collagen hydrolysate derived from bovine limed split wastes, *J. Soc. Leather Technol. Chem.* 90 (2005) 23–28.

62. J. Bella, A new method for describing the helical conformation of collagen: Dependence of the triple helical twist on amino acid sequence, *J. Struct. Biol.* 170 (2010) 377–391. doi:10.1016/j.jsb.2010.02.003.

63. A. Sionkowska, Current research on the blends of natural and synthetic polymers as new biomaterials: Review, *Prog. Polym. Sci.* 36 (2011) 1254–1276. doi:10.1016/j.progpolymsci.2011.05.003.

64. J.A. Fallas, L.E.R. O'Leary, J.D. Hartgerink, Synthetic collagen mimics: self assembly of homotrimers, heterotrimers and high order structures, *Chemical Society Reviews*, 39 (2010) 3510–3527. doi:10.1039/b919455j.

65. X. Hu, P. Cebe, A.S. Weiss, F. Omenetto, D.L. Kaplan, Protein-based composite materials, *Mater. Today.* 15 (2012) 208–215. doi:10.1016/S1369-7021(12)70091-3.

66. F. Nudelman, P.H.H. Bomans, A. George, G. de With, N.A.J.M. Sommaerdijk, The role of the amorphous phase on the biomimetic mineralization of collagen, *Faraday Discuss.* 159 (2012) 357–370. doi:10.1039/C2FD20062G.

67. J. Gosline, M. Lillie, E. Carrington, P. Guerette, C. Ortlepp, K. Savage, Elastic proteins: Biological roles and mechanical properties, *Phil. Trans. R. Soc. Lond. B.* 357 (2002) 121–132. doi:10.1098/rstb.2001.1022.

68. B.S. Kumar, S. Aigal, D.V. Ramesh, Air-dried 3D-collagen – Chitosan biocomposite scaffold for tissue engineering application, *Polym. Compos.* (2012) 2029–2035. doi:10.1002/pc.

69. J. Negishi, S. Funamoto, Y. Hashimoto, K. Yanagisawa, PLA-collagen composite scaffold fabrication by vacuum pressure impregnation, *Tissue Eng.* (2019) 1–18. doi:10.1089/ten.TEC.2019.0226.

70. M. Yeo, H. Lee, G. Kim, Three-dimensional hierarchical composite scaffolds consisting of polycaprolactone, β-tricalcium phosphate, and collagen nanofibers: Fabrication, physical properties, and in vitro cell activity for bone tissue regeneration, *Biomacromolecules* 12 (2011) 502–510.

71. L. Wei, H. Chang, Y. Wang, S. Hsu, L.-G. Dai, K.-Y. Fu, N.-T. Dai, A gelatin/collagen/polycaprolactone scaffold for skin regeneration, *PeerJ.* 7 (2019) 1–22. doi:10.7717/peerj.6358.

72. N. Munir, A. Callanan, Novel phase separated PCL/collagen scaffolds for cartilage tissue engineering, *Biomed. Mater.* (2018) 1–20. doi:https://doi.org/10.1088/1748-605X/aac91f Manuscript.

73. S. Kiran, K.C. Nune, R.D.K. Misra, The significance of grafting collagen on polycaprolactone composite scaffolds: Processing–structure–functional property relationship, *J. Biomed. Mater. Res. Part A* 103 (2015) 2919–2931. doi:10.1002/jbm.a.35431.

74. X. Yang, M. Yuan, W. Li, G. Zhang, Synthesis and properties of collagen/polylactic acid blends, *J. Appl. Polym. Sci.* 94 (2004) 1670–1675. doi:10.1002/app.21056.

75. C. Zhijiang, Y. Guang, Bacterial cellulose/collagen composite: Characterization and first evaluation of cytocompatibility, *J. Appl. Polym. Sci.* 120 (2011) 2938–2944. doi:10.1002/app.

76. A.P. Mathew, K. Oksman, D. Pierron, M.-F. Harmad, Crosslinked fibrous composites based on cellulose nanofibers and collagen with in situ pH induced fibrillation, *Cellulose* 19 (2012) 139–150. doi:10.1007/s10570-011-9624-x.

77. J. Audic, B. Chaufer, G. Daufin, Non-food applications of milk components and dairy co-products : A review, *Le Lait.* 83 (2003) 417–438. doi:10.1051/lait.

78. Wusigale, L. Liang, Y. Luo, Casein and pectin: Structures, interactions, and applications, *Trends Food Sci. Technol.* 97 (2020) 391–403. doi:10.1016/j.tifs.2020.01.027.

79. H. Chen, J. Wang, Y. Cheng, C. Wang, H. Liu, H. Bian, Y. Pan, J. Sun, W. Han, Application of protein-based films and coatings for food packaging: A review, *Polymers (Basel)* 11 (2019) 1–32.

80. J. Gu, J.M. Catchmark, Polylactic acid composites incorporating casein functionalized cellulose nanowhiskers, *J. Biol. Eng.* (2013) 1–23. doi:10.1186/1754-1611-7-31.

81. S. Zhang, X. Jin, X. Gu, C. Chen, H. Li, Z. Zhang, J. Sun, The preparation of fully bio-based flame retardant poly(lactic acid) composites containing casein, *J. Appl. Polym. Sci.* 46599 (2018) 1–9. doi:10.1002/app.46599.

82. J. Ma, Q. Xu, D. Gao, J. Zhou, J. Zhang, Blend composites of caprolactam-modified casein and waterborne polyurethane for film-forming binder: Miscibility, morphology and properties, *Polym. Degrad. Stab.* 97 (2012) 1545–1552. doi:10.1016/j.polymdegradstab.2012.04.015.

83. Y. Wang, J. Ma, Q. Xu, J. Zhang, Fabrication of antibacterial casein-based ZnO nanocomposite for flexible coatings, *Mater. Des.* 113 (2017) 240–245. doi:10.1016/j.matdes.2016.09.082.

Chapter 8

Surface Modification of Biobased Polysaccharide Nanoparticles via Grafting

R. Mincheva, S. Benali, and J.-M. Raquez
Center of Innovation and Research in Materials and Polymers (CIRMAP), Research Institute for Materials Science and Engineering, University of Mons-UMONS Place du Parc, 20, B-7000 Mons, Belgium
jean-marie.raquez@umons.ac.be

8.1 Introduction

Since the work of H. Staudinger in 1922 [1], no one can question the added value of macromolecular science on new manufacturing objects. In the same way, in the last few decades, nanocomposites have become an essential sector of polymer science. Indeed, particles 80,000 thinner than a human hair can be a very interesting alternative for achieving not only mechanical properties but also optical, thermal, gas barrier, and flame retardancy properties of polymers. In the earlier stages, it was mainly the so-called "hard" nanomaterials such as carbon or clays that have been used across

Green Micro- and Nanocomposites
Edited by Sabu Thomas, Abitha V. K., and Hanna J. Maria

ISBN 978-981-4968-79-9 (Hardcover), 978-1-003-42756-8 (eBook)
www.jennystanford.com

many different applications with more or less success [2]. In another stream of activities, always driven by health and environmental concerns, biodegradable crystalline nanoparticles from renewable resources have attracted significant interest from the scientific community. In this framework, carbohydrate-based (polysaccharide) nanofillers have gained increased interest as part of an effort to avoid nanotoxicity and overcome environmental concerns [3].

The family of polysaccharides is characterized by a common morphology composed of alternating crystalline and amorphous zones. The latter are highly susceptible to hydrolysis under defined conditions, with their removal leaving the crystalline regions intact in the form of remarkable nanoparticles – polysaccharide nanocrystals [4]. Whether from an environmental point of view or for specific applications, polysaccharide nanocrystals offer multiple promising opportunities [4], in particular, in drug delivery systems [5], nanoencapsulation [6], or as reinforcing agents in polymer matrices [7–12]. With regard to the potential of these natural, available, and low-cost raw materials, the thinking around new application fields will contribute to a greener and more sustainable technology development. Thorough bibliographic studies mainly point to cellulose nanocrystals, as they are characterized by high structural stability under chemical modification in comparison with chitin or starch nanocrystals. However, chitin nanowhiskers and starch nanoplatelets also deserve full attention for promoting specific applications based on their unique properties. Nowadays, industrial opportunities for natural polysaccharide nanocrystals are under development around the world. As an example, industrial production of nanocelluloses is promoted by an increasing number of countries supporting their valorization. Here, the construction of multiton capacity manufacturing facilities in North America, Japan, and Europe can be cited [4, 13–16].

Despite all benefits, the applicability of the polysaccharide nanoparticles faces significant difficulties resulting from their inherent hydrophilicity – the basis for incompatibility and bad dispersibility within most polymer matrixes (hydrophobic in character). In order to avoid these limitations, after appropriate conversion and extraction technologies, and suitable modification, polysaccharide nanocrystals might find a place in numerous biobased products. Initially, the development of polysaccharide-reinforced nanocomposites was mainly restricted to hydrophilic media as these nanocharges possess inherent hydrophilicity and aggregate in hydrophobic media. However, their surface modification (mostly chemical) overcomes this limitation and has been largely applied in broadening the number of suitable polymeric matrices with processing from organic dispersions or melts instead of aqueous suspensions. In this light, promising developments in the design process of nanocelluloses allow to focus on research on all polysaccharide nanocrystals, including chitin and starch, with particular attention on their surface modification for better compatibilization with biobased materials.

Amongst all methods, surface grafting of polymer chains is a promising and attractive new process by which polysaccharide nanoparticles are transformed into apolar hybrid nanoparticles. The procedure is readily intriguing as fully biobased composites can be processed by classical methods such as extrusion, compression or injection molding, or thermoforming [17].

Many interesting and progressive works have been performed on the surface modification of biobased polysaccharide nanoparticles via grafting, thus paving the road for the upcoming chapter.

8.2 Surface Grafting

In terms of chemistry, grafting is an often-used modification method through which polymers are introduced to a surface [18]. It allows for modulation of surface properties in terms of hydrophilicity/hydrophobicity, dimension stability, resistance to heat, abrasion and/or chemicals, biological activity or electrical properties, and functionality [18]. Depending on how polymers are introduced, three major types of grafting can be defined: 1) grafting from, 2) grafting onto, and 3) grafting through (Fig. 8.1). However, grafting can also be performed via a non-covalent approach – the so-called supramolecular grafting (Fig. 8.1d). This new and yet not much explored trend uses softer conditions and prevents surface integrity and functionality. The following parts discuss the different grafting approaches in terms of requirements and applicability to solid surfaces.

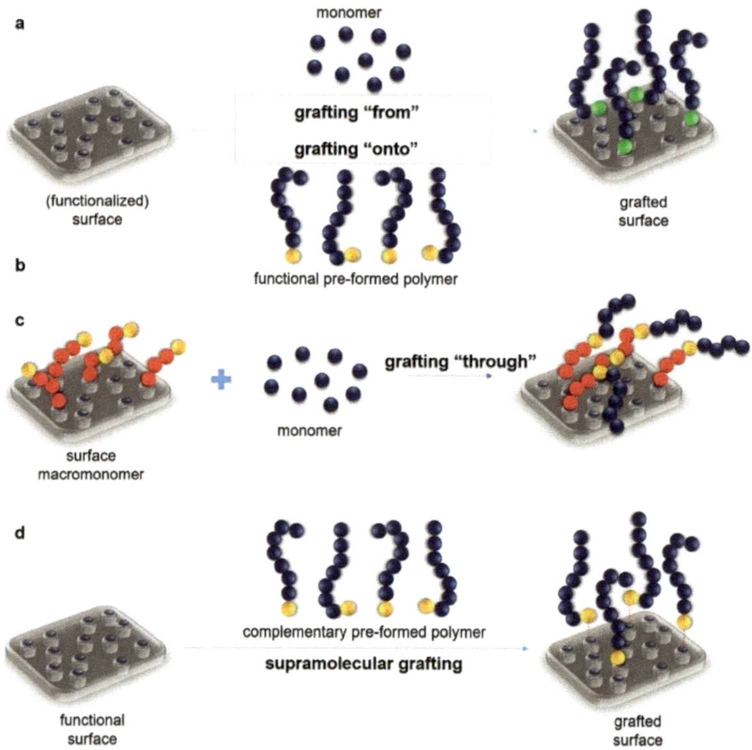

Figure 8.1 Schematic representation of the grafting methods used for introducing polymer chains to a surface.

8.2.1 Grafting From

The *grafting from* approach includes the *in situ* building up of a polymer layer by surface-initiated polymerization of monomer units (Fig. 8.1a). For this, an appropriate initiator (initiating sites) is first introduced on the surface to be grafted via external field (plasma, UV irradiation, ozone treatment) or reactive species introduction (chemical modification) [19]. Intriguingly, in some cases, the initiating sites are naturally available on the surface via the presence of reactive groups (e.g., OH-groups on cellulose substrates). A very attractive peculiarity here is the fact that the initiating sites are usually easy to access by the monomer molecules forming the chain ends of the covalently attached growing polymers. Thus high grafting densities are achieved making this method very attractive from a scientific point of view [19].

Multiple polymerization methods such as surface-initiated free radical polymerization (SI-FRP), controlled radical polymerization (SI-CRP), and ring-opening polymerization (SI-ROP) have been elaborated in grafting *from* [20]. Like conventional FRP, the SI-FRP is a versatile chain-growth polymerization applicable to a large range of monomers, thus providing an unlimited number of (co)polymers [21]. The tolerance of this method to water or other impurities and functionalities (e.g., OH, OR, NH_2, NR_2, etc.), together with the mild and versatile (solution, emulsion, bulk, etc.) reaction conditions, makes the SI-FRP highly attractive. The method comprises all reaction steps of the conventional FRP (Fig. 8.2) and allows for obtaining a high grafting density due to the ease of contact between SI sites and monomers [18, 22, 23].

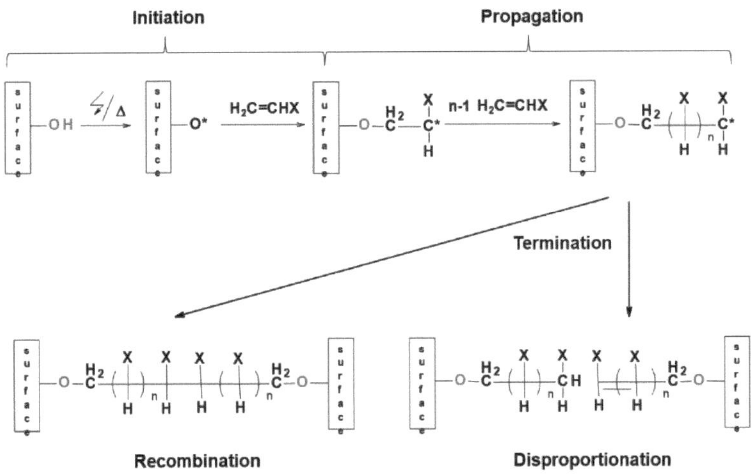

Figure 8.2 Schematic representation of the general FRP *grafting from* mechanism.

However, difficulties in the correct control or prediction of grafting density and resulting polymer chain length and architecture are often found. Moreover, the formation of free, unbound homopolymers is unavoidable. Therefore, other research groups introduced the SI-CRP techniques to the *grafting from* method [20].

Amongst all, the surface-initiated CRPs, nitroxide-mediated polymerization (NMP) [24], atom transfer radical polymerization (ATRP) [25–27], photoinduced Cu-mediated reversible-deactivation radical polymerization (RDRP) [22], and reversible addition–fragmentation chain transfer polymerization (RAFT) [28] received greatest attention [18]. Their principles (Fig. 8.3) are exhaustively discussed elsewhere [18, 20, 29] and therefore will only be recalled here. The NMP belongs to the family of stable RPs and involves a stable nitroxide radical (X^*) [30]. Here, the active propagating species (P_n^*) react with X^*, deactivate, and reversibly form dormant species (P_n–X). Once P_n^* reforms, it re-propagates by adding another monomer (M) or terminates the chain growth by recombination (Fig. 8.3A). NMP is usually performed under relatively high (> 100 °C) temperatures and is applicable to acrylate-type monomers. However, its application in methacrylates/methacrylamides requires specially designed stable radicals.

In a reversed way, ATRP starts with an alkyl halide (P_n-X) that undergoes a reversible redox process catalyzed by a transition metal complex (activator, Me^z-Y/ligand, where Y may differ from the ligand or be a counter ion) and forms the P_n^* and a metal halide complex (X···Me^{z+1}–Y/L) [31]. The P_n^* reacts with a monomer and propagates or abstracts a halide atom from X···Me^{z+1}–Y/L and reforms dormant species. The alkyl halide species is then reactivated by the activator and propagate further (Fig. 8.3B). ATRP is easily applicable to most of the vinyl and acrylic monomers over a wide temperature range and is known to be tolerant (to a certain extent) to oxygen and other inhibitors. Nevertheless, the reaction uses unconventional initiation systems that may suffer from poor solubility in the polymerization media. Additionally, the transition metal residues color and induce certain toxicity to the obtained polymers.

RAFT is the polymerization method of choice when grafting of vinyl and acrylic monomers is desired [18]. This polymerization is similar to the conventional RPs when performed in presence of a chain transfer agent ("RAFT agent" – a thiocarbonyl thio compound). It results in polymers of narrow dispersities and of controlled chain lengths (Fig. 8.3C). The problem that one can relate to the coloration of the resulting product by the thiocarbonyl thiol end-group is easily solved by reduction, thermolysis, aminolysis, exposure to ultraviolet radiation, and treatment with peroxides or sodium hypochlorite.

Thus, it seems the RAFT process is the method of choice for grafting from cellulose.

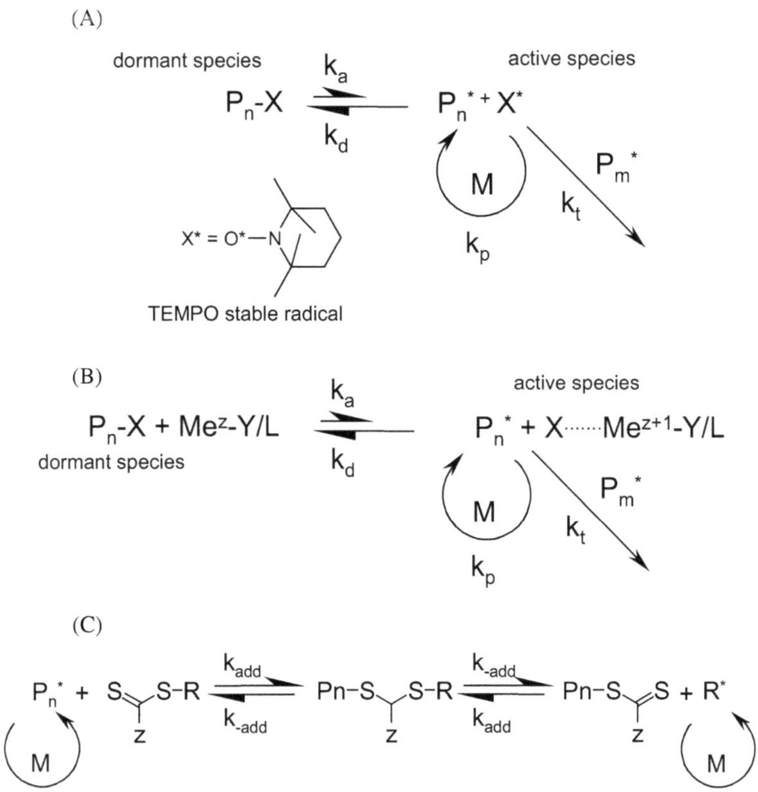

Figure 8.3 Schematic representation of the accepted mechanisms for NMP (A, [30]), ATRP (B, [31]), and RAFT (C, [18]).

However, these methods are not applicable in the case of cyclic monomers (lactones and lactides), where ROP is commonly used. This well-established technique relies upon alcohol (in general) initiation, which makes it suitable in cellulose or cellulose derivatives grafting from [32]. Depending on the monomer, initiator, and catalyst used, ROP operates through different mechanisms. For example, in the case of tin(II) 2-ethylhexanoate ($Sn(Oct)_2$) catalyzed ROP of monomers such as ε-caprolactone (ε-CL), lactide (LA), and p-dioxanone, the most commonly accepted mechanism is the "coordination-insertion" mechanism in which $Sn(Oct)_2$ converts to tin alkoxide (the actual initiator) by reaction with alcohols or other protic compounds/impurities. The fine-tuning of the alcohol-to-monomer ratio allows better control over the molecular weight of the final polymer [32].

Nevertheless, none of the methods discussed here allows for overcoming the formation of non-grafted polymer chains or knowing their molecular characteristics. As a result, grafting onto is usually applied.

8.2.2 Grafting Onto

In grafting onto (Fig. 8.1b), usually end-functionalized preformed polymer chains of known molecular characteristics (chemical structure, molecular weight, dispersity, morphology, etc.) are covalently attached to a (modified)hydroxyl groups of a substrate. Additionally, the grafting might be performed via the deactivation of living polymer chain ends by the surface functional groups, thus allowing attaching commercial polymers with a well-characterized structure, which makes the process very attractive for industrial applications [33]. For the method to be successful, it is necessary to apply efficient coupling chemistry such as 1) reactions between a living polymer chain end-group and a suitable group on the surface to be grafted [34, 35], 2) copper(I) catalyzed Huisgen 1,3-dipolar cycloaddition [36–38], 3) esterification and amidation [39–43], 4) isocyanate chemistry [17, 44, 45], and 5) nucleophilic substitution [46, 47] (Fig. 8.4).

However, steric hindrance can prevent reaching the available reactive sites at the surface, thus unfavorably reducing the surface grafting density via the grafting onto the method. A method aiming on combining grafting onto and grafting from to overcome their inconvenience is the grafting through.

8.2.3 Grafting Through

The grafting through process (Fig. 8.1c) involves (co)polymerization of macromonomer(s) [18, 30]. For its purposes, the surface to be grafted is firstly functionalized with a polymerizable group (a vinyl group) able to participate in polymerization as a macro-monomer (Fig. 8.5) [49]. As a particularity of the method, there is no relation between the amount of nanoparticles and initiator concentration.

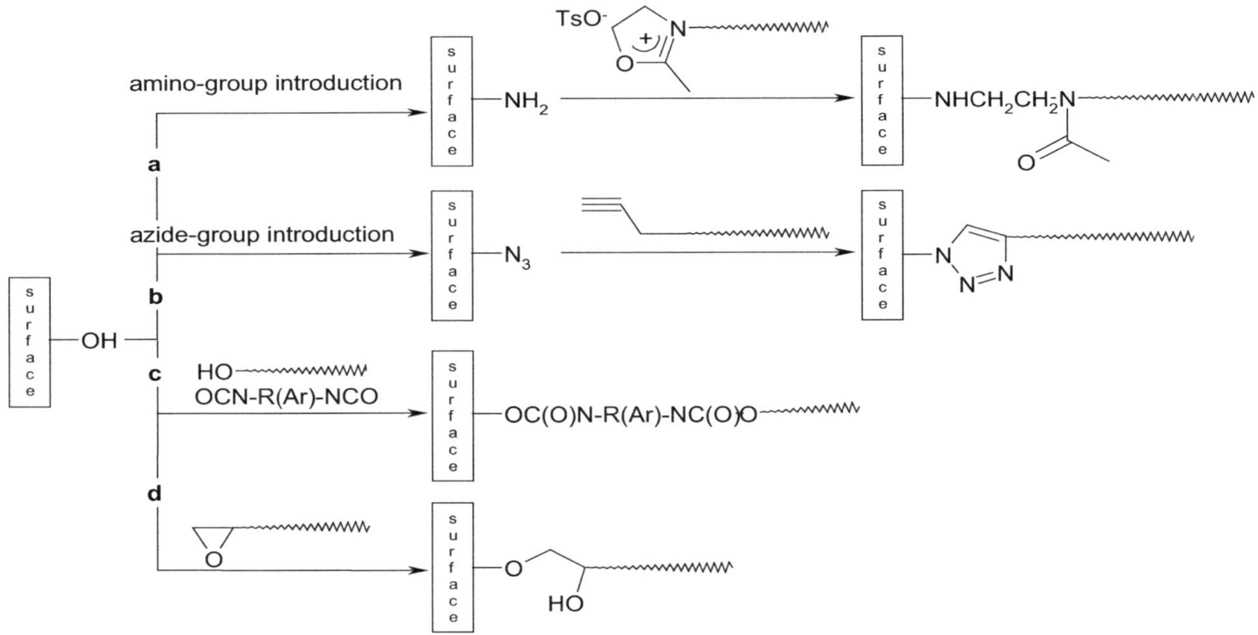

Figure 8.4 Grafting *onto* nanosurfaces: a – reactions between a living polymer chain end-group and a suitable group [35], b – copper(I) catalyzed Huisgen 1,3-dipolar cycloaddition [48], c – isocyanate chemistry [38], and d – nucleophilic substitution [46].

The grafting through technique overcomes the growth of long chains and promotes this of shorter chains. This is achieved by reversing the monomer concentration gradient in the grafting-from, where the monomer concentration is lowest at the substrate and highest in the surrounding solution.

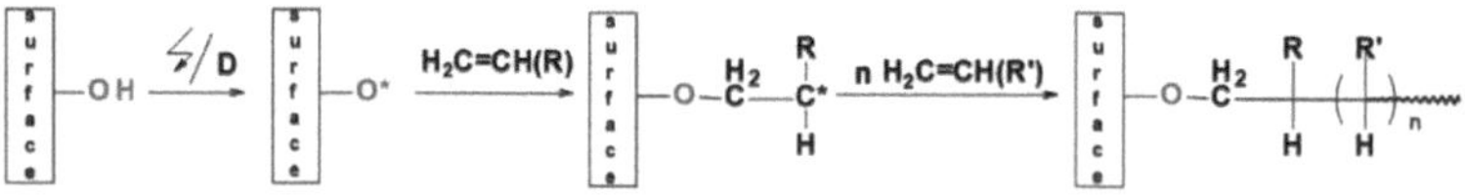

Figure 8.5 Grafting through CNC macromonomers.

However, the grafting through is yet scarcely applied due to the following reasons:

(1) Limited availability of polysaccharide macromonomers
(2) The possible interreference of the surface chemical modification for introducing suitable functionalities with polysaccharide nanocrystal (PSNC) integrity, crystallinity, and other surface properties [18]
(3) Possibility to cross-link the system via recombination of growing chains with two or more propagating sites on macromonomers [50]

Thus, further research must be done for taking advantage of the grafting through approach, or, alternatively, supramolecular grafting can be used.

8.2.4 Supramolecular Grafting

Supramolecular chemistry is non-covalent chemistry employing hydrogen bonding, metal coordination, hydrophobic forces, van der Waals forces, pi-pi interactions, and electrostatic effects in molecular self-assembly, folding, and recognition, host-guest chemistry, mechanically interlocked molecular architectures, and dynamic chemistry [51]. For several years, the method has been an alternative of choice for the surface modification of nanoparticles mainly because it involves milder and nondestructive conditions for surface functionalization/modification/grafting and thereby preserves structure, properties, and straightforwardness of nanoparticles [52]. When supramolecular chemistry is used for surface grafting, the so-called supramolecular grafting, preformed polymer chains of known molecular characteristics and having special functionalities along their backbones react with the available functional groups of the surface to be modified in a mild and complementary nondestructive manner (Figs. 8.1d and 8.6). Thus, the problems related to grafting from technique are avoided. Besides, the possibility to design and incorporate suitable functionalities along the preformed polymer chains overcomes the need for surface functional group modification.

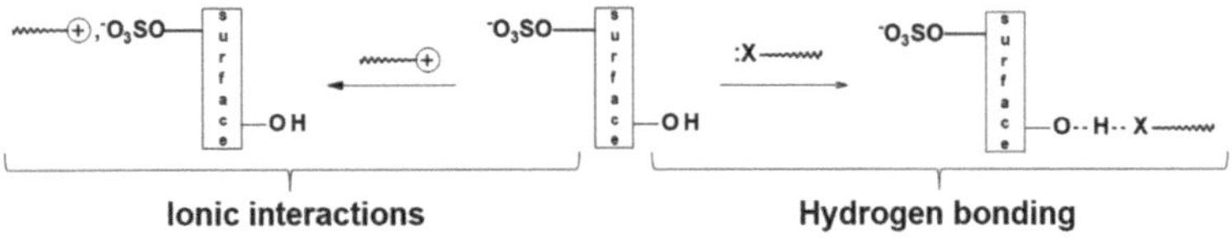

Figure 8.6 Chemistry of some supramolecular interactions.

Multiple examples of the application of these grafting techniques to polysaccharide nanoparticles (cellulose, starch, and chitin) will be given below, but prior to this, the next section discusses the particular base for their applicability.

8.3 Polysaccharide Nanoparticles

8.3.1 Raw Material

Polysaccharides nanocrystals can be "extracted" from raw materials such as cellulose and chitin (the most abundant polymers on Earth) or from starch (the major energy reserve of higher plants) and offer molecular and biological advantages for their use in the preparation of nanocomposites.

Cellulose is extracted from the cell walls of plants, from algae, tunicates, and bacteria [53, 54]. The exoskeleton of crustaceans, shellfish, and insects is the principal source of chitin [55] and starch can be found in seeds such as wheat, corn, or rice or in tubers such as potatoes [7, 56–58]. The chemistry of these polysaccharides varies from linear β–(1→4)-linked D-glucose residues (cellulose, Fig. 8.7A) to linear *N*-acetyl-D-glucosamine units linked through β–(1→4)-glycosidic linkage (chitin, Fig. 8.7B); or a mixture of amylose, a linear or slightly branched (1→4)-α-D-glucan, and amylopectin, a highly branched macromolecule consisting of (1→4)-α-D-glucan short chains linked through α-(1→6) linkages (starch, Fig. 8.7C). These carbohydrates are widely used in food (starch), clothing (cotton), communication (paper), packaging (paper and board), and construction (wood) and have always been a fundamental part of both industrial and academic R&D [59]. Furthermore, due to the depletion of fossil resources, climate changes, and toxicity impacts, polysaccharides remain pivotal options for the future of biobased materials. Currently, interest in polysaccharides has shifted toward nanoscale materials.

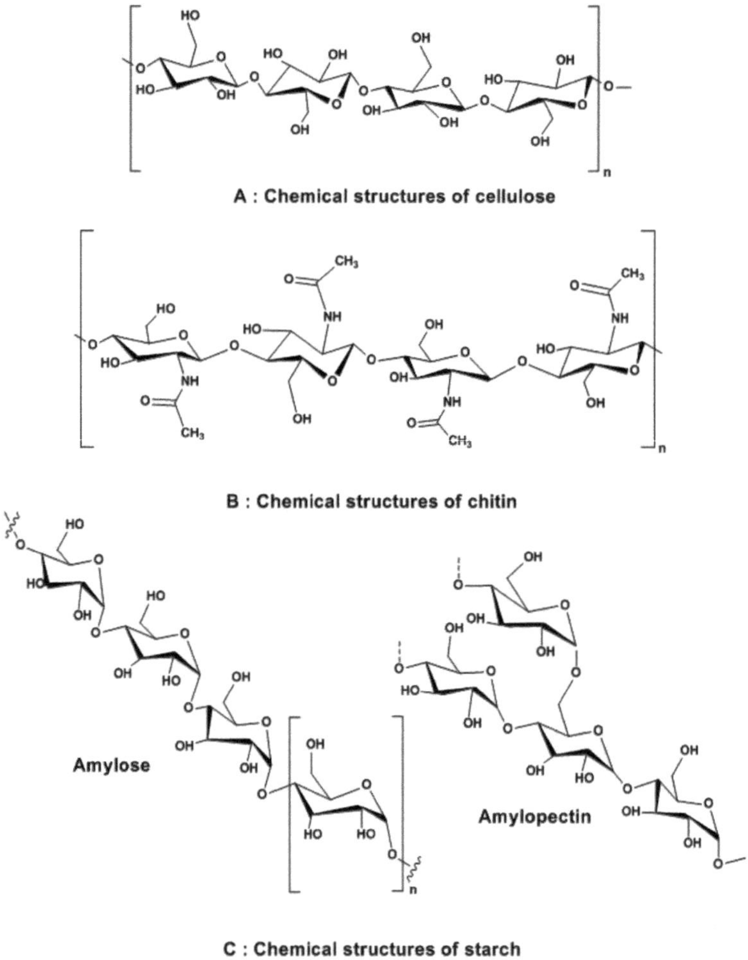

Figure 8.7 Chemical structures of polysaccharides.

However, the greatest obstacles to their expansion are (i) the low thermal stability of polysaccharide nanocrystals to consider melt processing and (ii) poor compatibility with a number of polymer matrices. In this framework, surface modification of polysaccharide nanocrystals appears to be a powerful strategy to meet these challenges.

8.3.2 Preparation

8.3.2.1 Nanocellulose

Over the past several decades, there has been extensive research in nanocellulose. Nickerson and Habrle first reported an efficient procedure to prepare nanocellulose from cotton linters [60]. In 2011, the Technical Association of the Pulp and Paper Industry (TAPPI) confronted with numerous studies about cellulose nanocrystals and considered it appropriate to bring together experts to release a harmonized nomenclature about nanocellulose materials [61]. Based on their recommendations, several reviews and books described the methods for extracting different cellulose nano-objects, such as cellulose nanofibrils (CNFs) and cellulose nanocrystals (CNCs) [4, 12, 13, 54, 62–65] from plants (e.g., cotton [66], hemp [67], wood [68]), marine animals (e.g., tunicates [69]), algae (e.g., *Valonia* [70]), bacteria (e.g., *Acetobacter xylinum* [71]), and even amoeba (*Dictyostelium discoideum* [72]) to fabricate a wide range of functional materials such as reinforcing filler [54] as well as photonic crystal [73], barrier film [74], shape-memory polymers [75], light-healable [76], drug-delivery [77] and mechanically adaptive nanocomposites [78]. There is a significant difference to be understood between CNF and CNC, although one is coming from the other. CNFs are bulky products of nanocellulose production with a large surface area and aspect ratio: diameters ranging from 20 to 60 nm and lengths of several microns. They are usually the finest result of the mechanical treatment stage, without any acid contact. While the terms microfibril, nanofibril, nanofiber, and elementary fibrils are usually used as synonyms for CNFs, they must be avoided in describing CNC – the finest product of acid hydrolysis of CNF amorphous regions. In contrast to CNFs, CNCs are rodlike particles of low aspect ratio and an elastic modulus of about 140 GPa [13, 79].

As the methods for nanocellulose production are well described in numerous studies, reviews, and chapters, they will not be discussed here. However, Fig. 8.8 summarizes cellulose fiber treatments for extracting both CNFs and CNCs, and some additional explanations are given below.

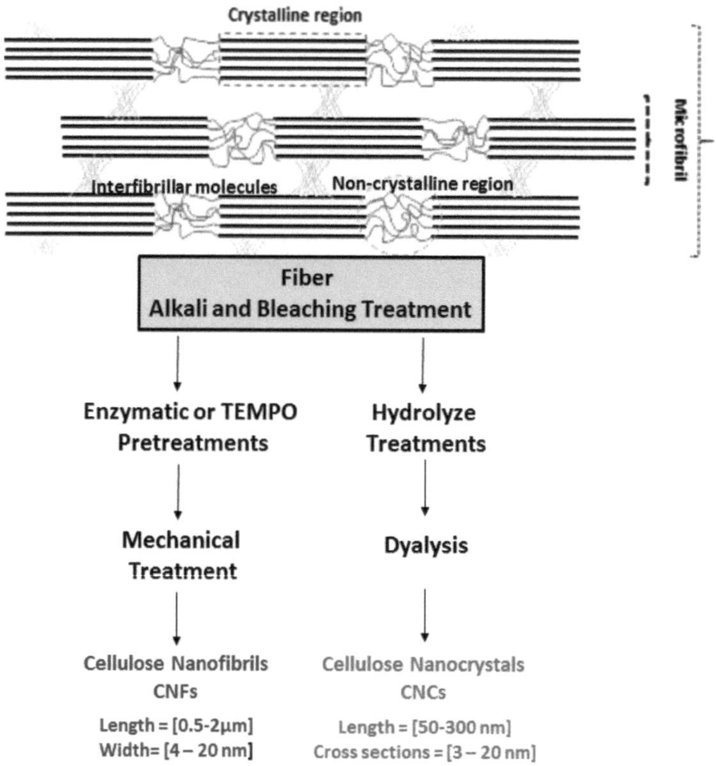

Figure 8.8 Schematic representation of the extraction procedures for both CNFs and CNCs from cellulose.

With respect to cellulose nanofibrils (CNFs), scaling-up production and reduction of energy consumption forced researchers to set up a global procedure including chemical or enzymatic pretreatments before a mechanical treatment and high-pressure homogenization process [13, 53, 54, 80–82]. Regarding the procedure to extract cellulose nanocrystal (CNCs), ideally the amorphous regions are hydrolyzed and the remaining crystals will be nanometer size [13]. Actually, the shape of CNCs is strongly dependent on acid hydrolysis conditions. Composed of β–(1→4)-linked D-glucose units, linear polymer chains are arranged in highly crystalline cellulose I (native cellulose). The typical methodology for CNC extraction involves strong-acid hydrolysis under well controlled conditions (temperature, stirring, and time) of purified cellulose. Depending on the source and parameters of the hydrolyze, the author's product ranges from 50 to 300 nm in length and cross sections of 3–20 nm [4, 12, 13, 53, 62]. A comparative and thorough reading of the literature makes it possible to show that the structure, properties, and behavior of the CNCs are strongly dependent on their source and the extraction method [4, 13, 62]. Additionally, the production (as integrated forest biorefinery [80]) changed scale, forcing the users to adequately characterize nanofillers prior to any use. The real question, nowadays, is whether production scale, starting material, or purification affects nanocellulose behavior. A recent publication by Reid et al. [16] proposes an interesting benchmarking of CNCs from laboratory to industrial production. Overall, sulfuric acid extracted CNCs compare well with CNCs produced at the laboratory scale with the final product being highly crystalline, high aspect ratio "nano-only" CNCs. However, differences in sulfate half ester content, colloidal stability, crystallinity, and morphology are clearly observable. It is established that the community needs to have industrially produced nanocelluloses to continue to develop nanocellulose research. These results force both science and industry to set up proper characterization to forecast nanocellulose behavior [16].

8.3.2.2 Nanochitin

The fully acetylated version of chitin is insoluble in water and in most organic solvents due to the strong hydrogen bonding between the acetyl group. In this light, chitin is challenging to work, and it is difficult to maintain the chitin nanofiber structure during processing. Chitin exists in three polymorphs α, β, and γ that differ in orientation and packing of the chitin molecular chains. The most abundant is the α-chitin (*from* crab and shrimp shells). There, the polymer chains align in an antiparallel arrangement that favors strong intermolecular hydrogen bonds. In contrast, the β-chitin (*from* squid pens) polymer chains pack in parallel, thus weaking intermolecular interactions. The γ-chitin structure is a mixture of the α- and β-forms [55]. A very recent publication investigating engineering strategies for chitin nanofibers (ChNF) allows to review all previously developed syntheses by both "top-down" and "bottom-up" routes [83]. So far, the top-down approach, consisting of breaking the native chitin microfibril to individual building blocks of interest, has been preferred in order to prevent the difficulties of dissolution at the molecular level. In this approach, pure chitin is first isolated upon demineralization and deproteinization using acid and alkali treatment. Then, the resulting mass is strongly mechanically disintegrated to chitin nanofiber dispersion with range diameters of 10–20 nm [83–87]. Wu et al. proposed an easier route to successfully extract ChNF from crab α-chitin by a milder high-pressure homogenization process under acidic conditions [88]. The synergistic effect between high-pressure homogenization and cationization of chitin was also found to disintegrate effectively the large chitin fibers to give nanofibers of an average diameter of 20 nm [88]. Other 'top-down' approaches are mediated oxidation with strong acid hydrolysis, intensive mechanical disintegration, and 2,2,6,6-tetramethylpiperidine-1-oxyl radical (TEMPO), which led to ChNF with an average width of 8 nm [83, 89] or a chemical-etching-free approach to disintegrate chitin using calcium ions and solvent exchange [90]. In parallel, the bottom-up approach involves dissolving chitin fibers before reassembling into chitin nanofibers. Some microfibers have been prepared using electrospinning after dissolution into 1,1,1,3,3,3-hexafluoro-2-propanol (HFIP) with an average diameter of 160 nm. A pretreatment with ionic liquids allows dissolving chitin before electrospinning for making thinner microfibers (i.e., average diameter between 30 and 80 μm). Zhang and Rolandi proposed a different route of self-assembling chitin nanofibers from squid pen β-chitin previously dissolved into HFIP or LiCl/*N*,*N*-dimethylacetamide (DMAC) with electrospinning free step [83]. Promising chitin nanofibers with a diameter of 3 nm have been obtained.

The procedures for obtaining chitin nanocrystals (ChNCs) are obviously very similar to the one for CNCs [91] and will not be a subject of this review.

8.3.2.3 Nanostarch

Introduced by Le Corre et al. [7], Lin et al. [4, 57], or, more recently, Kim et al. [92, 93], the main method to obtain starch nanocrystals relies again on the disruption of the amorphous and paracrystalline domains of semicrystalline starch granules via acid hydrolysis (Fig. 8.9). The optimization of the process is studied by varying diverse parameters including the acid used or the botanic origin of the native starch [94, 95]. However, the first studies of Le Corre et al. [7], Lin et al. [57], or, more recently, Kim et al. [93] are the basis of a global review on the preparation of starch nanocrystals with lengths ranging 20–150 nm with possible crystallinity preservation. In parallel, starch nanoparticles were prepared from gelatinized starch using numerous physical treatments including extrusion [96], high-pressure homogenization and emulsification [56, 97], ionic liquid medium [98–100], solvent displacement method [58, 93, 101], or ultrasonication [102] with promising results for industrial applications [93].

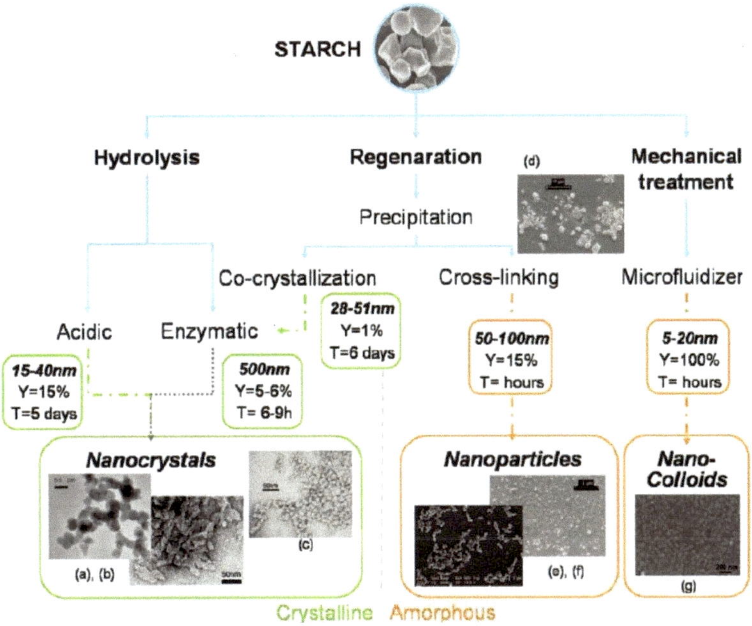

Figure 8.9 Different ways of producing crystalline and amorphous starch nanoparticles: hydrolysis leads to nanocrystals, whereas regeneration and mechanical treatment lead to both amorphous and crystalline particles in the final batch. Adapted with permissions from Ref. [7]. Copyright 2019 American Chemical Society.

For example, Song et al. investigated the mechanism of starch nanoparticle formation during extrusion. They show that an appropriate use of a cross-linker at 75°C facilitates the reduction of particle size to around 160 nm [96]. However, as expected, mechanical damages are caused by the longtime of melt blending under high shear. On the other side, with the high-pressure homogenization technique, Liu et al. developed an alternative sustainable approach to reducing the size of SNPs. This method uses a microfluidizer to manipulate the continuous flow of liquid through microchannels [93]. Based on specific experimental parameters (starch slurry, 20 passes, high pressure), the size of starch granule particles decreases from a few micrometers to a few nanometers. However, experimental conditions must be optimized for better preservation of the crystalline structure and improving the recovery yield [97]. The nanoprecipitation process is also a promising anti-solvent method to extract SNPs. This method consists of the deposition of polymers using a semipolar solvent miscible with water from a hydrophobic solution [92]. Here, the successive addition of a diluted starch solution to a non-solvent leads to the precipitation of nanostarch [58]. In this respect, the nanoprecipitation method, initially reported by Qin et al., allows to obtain differential structural and morphological properties of SNPs using several native starches with various amylose content and different types of crystalline structures. Authors conclude that the obtained SNPs display a typical V-crystalline structure (i.e., nanoplatelet-like) with particle sizes ranging from 20 to 225 nm depending on native starch granules.

8.3.3 Physicochemical Properties

Physicochemical properties of polysaccharides nanocrystals have been outlined clearly by Lin et al. [4]. In the framework of this chapter, Table 8.1 proposes a comparative presentation of the main differences between cellulose, chitin, and starch nanoparticles. In general, the morphology and geometrical dimensions of polysaccharide nanocrystals are related to the starch origins and extracting method. More specifically, common cellulose nanocrystals with rodlike morphology derived from cotton, flax, ramie, sisal, and so on are characterized by a length range of 100–700 nm and diameter 5–30 nm, while animal tunicate presents a considerable aspect ratio of about 100 and bacterial cellulose gives L and D of 100 to several mm and from 5 to 50 nm, respectively [91]. Dimensions of chitin nanowhiskers extracted from shrimp shell, crab shell, or squid pen were found to be close to those reported for cotton whiskers, while for Riftia tubes ChNC, the average L was around 2.2 mm and the aspect ratio was 120 [4]. As for the platelet-like starch nanocrystals, they are generally derived from crops, such as pea, potato, corn, and waxy maize L is between 20 and 100 nm, W - around 25–30 nm, and T - 6–8 nm [4].

Table 8.1 Physical characteristics of polysaccharide nanocrystals

NC	Morphology	Stiffness, GPa	Crystallinity, %	T_m, °C	OH content, mol/g
CNC	Rodlike [54] Spherical [103]	120–170 [104]	54–88 [63]	200–300 [105]	0.0038 [4]
ChNC	Rodlike [106] Whiskers [107]	150 [108]	>80 [109]		
SNP	Platelet-like [110]		40–50 [4]	250 [4]	0.0025 [4]

The small dimensions cause some difficulties in manipulating individual nanocrystals, and direct measurement of their stiffness is not easy. However, some data are reported from theoretical calculations, XRD analyses, or Raman spectra, well summarized in Table 8.1 and in a recent review [4].

For what is to the degree of crystallinity, theoretically, it should be total, but the often-incomplete removal of the amorphous phase results in lower values. The classical values reported in the literature are within the range 54–88% [63] for CNCs, >80% for ChNCs [109], and 45–50% for SNPs [57] depending on the sources.

8.3.4 Chemical Properties

From a chemical point of view, polysaccharide nanocrystals are a challenging platform possessing reactive surface hydroxyl (and amino) groups, which allow the modification using chemical reactions (Fig. 8.7). In the last decades, numerous expert studies focused on this possibility usually to ensure nanocrystals applications as strengthening agents in composite materials or specific properties for novel nanomaterials [4]. More precisely, in the case of rodlike cellulose nanocrystals with very uniform geometrical dimensions, around 0.0038 mol/g of active hydroxyl groups on the surface of nanocrystals is calculated [4]. Using the same calculations, the amount of surface available OH-groups in SNP was estimated to be ca. 0.0025 mol/g of the total amount [111].

Another significant point is the difference in the activity of the hydroxyl groups. The methylene OH group in the C6 position, for example, is known to be more active than the other two hydroxyl groups (C2 and C3). Besides, the chemical modification initially occurs at the surface of the nanocrystals, inducing a superficial chemical reaction [4]. Consequently, the reactive activities of the hydroxyl groups on the surface of polysaccharide nanocrystals can be quantified with the experiments as well as their different orders or gradient chemical modifications can be controlled.

Special attention should, however, be paid to reaction time, as with increasing it, surface modification can propagate to the inner part of the crystallites and induce their erosion and, subsequently, the loss of crystallinity [4]. Accordingly, an effort must be focused on the preservation of the structure, morphology, and crystalline properties of nanocrystals even though the degree of substitution or grafting efficiency is enhanced.

Attention is also needed to the amount of sulfur groups on the surface of the particles. Indeed, sulfuric acid is used as the hydrolyzing agent and hydrolysis treatment leads to the introduction of negatively charged sulfur groups ($OSO3^-/H^+$). The presence of sulfur groups favorises a homogeneous dispersion while keeping thermal stability due to the stabilization electrostatic of nanocrystals in water [4]. In the meantime, the decrease of potential surface hydroxyl groups, because of the incomplete replacement by the surface acid groups, reduces the reactivity for chemical modification. The same issues concern reactions with the amine functional groups available on the surface of chitin nanocrystals.

8.3.5 Surface Modification of Biobased Polysaccharide Nanoparticles via Grafting

All biobased polymer composites (biocomposites) where both the polymer and the (nano)particles (nanofillers) are of (biodegradable) renewable materials have received a tremendous interest over the last decades. Amongst all biofillers, polysaccharide nanocrystals including cellulose, starch, and chitin show superior properties as reinforcing reagents for biocomposites due to their biocompatibility, nontoxicity, and relatively low cost [112]. Unfortunately, hydrophilic nanocrystals self-aggregate easily, which leads to a low degree of dispersion and low efficiency of reinforcement in nanocomposites. Additionally, the hydroxyl groups on the surface of nanocrystals are immiscible with hydrophobic polymers as polyesters. To overcome this drawback, modification of the crystal surface is most commonly used, between all methods, via grafting techniques (Fig. 8.1). The applicability of each grafting method to all three types of polysaccharide nanoparticles will be discussed below.

8.3.5.1 The nanocellulose: nanocrystals and nanofibers

As already discussed in the previous section, nanocellulose is one of the most abundant and promising renewable nanomaterials for multiple applications that combines low thermal expansion, excellent mechanical properties, and high surface area with versatile modification capacity [113]. According to the Technical Association of the Pulp and Paper Industry (TAPPI), there are three forms of nanocellulose: cellulose microfibrils, cellulose nanofibrils, and cellulose nanocrystals, mainly based on variations in dimensions and flexibility [114]. A lot of research has been concentrated on the surface modification of these nanocelluloses and more particularly of the cellulose nanocrystals via grafting reactions. Most of the published papers are thoroughly summarized in several exhaustive reviews and books and therefore will not be discussed in detail here [12, 20, 29, 115]. Here, some new trends, mostly in relation to *supramolecular grafting*, will be discussed.

Nanocrystalline cellulose supramolecular interactions are well known and described in a multitude of studies [116]. However, supramolecular grafting is a new, softer, and nondestructive trend that in most cases does not need modification of the existing functional groups [52]. In the case of nanocelluloses, this grafting method was pioneered by the studies of Zhao et al. [117] who reported covalent modification of the cellulose microfibrils functional groups with cyclodextrins (CD) followed by *supramolecular grafting* of adamantine-capped poly(ε-caprolactone) (PCL) oligomers via the host-guest inclusion complexation in DMF dispersion (Fig. 8.10).

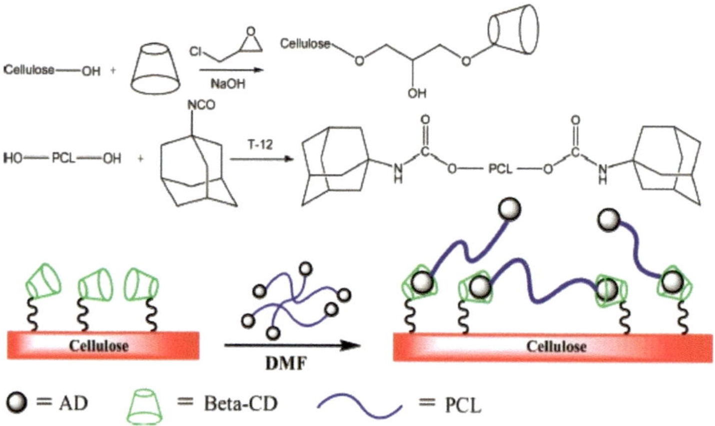

Figure 8.10 Synthesis pathway for the cellulose-CD and PCL-AD and the conceptual illustration for the assembly process of cellulose-CD with guest polymer PCL-AD (Reprinted with permission from [117]. Copyright 2017 American Chemical Society).

The successful assembly was confirmed by FTIR-ATR, XPS, and the increasing weight with CD concentration. Contact angle and TGA measurements reflected enhanced hydrophobicity and thermal stability of the cellulose fibers. Their morphological evaluations with SEM showed smooth surfaces exhibiting visible undulations along the axial direction, similarly to the pure cellulose before grafting [117]. Indeed, in this particular case, modification of cellulose functional groups prior to grafting was used but proved to be unnecessary by the following studies.

The supramolecular approach toward surface grafting of nanocelluloses was continued by the studies of Tatsumi et al. [118] in their attempts to synthesize novel composites comprising poly(2-hydroxyethyl methacrylate) (PHEMA) and cellulose nanocrystals (CNC) from CNC suspensions in aqueous 2-hydroxyethyl methacrylate (HEMA) monomer solution. The starting suspensions separated in isotropic upper and anisotropic bottom phases resulting after drying in transparent birefringent films of isotropic phase, embryonic non-separating mixture, and anisotropic phase, respectively. A fingerprint texture was found depending on the phase formed and the corresponding presence/absence of liquid-crystalline organization (Fig. 8.11) [118].

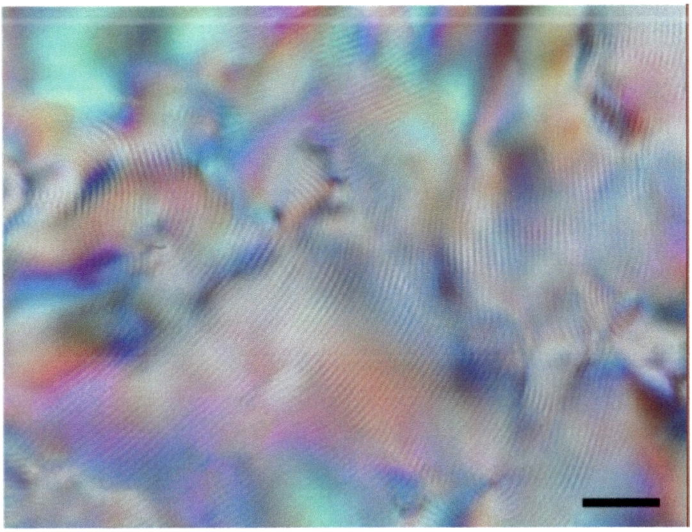

Figure 8.11 Polarized optical micrograph of the anisotropic phase of a 5.0 wt% CNC suspension in water/HEMA (0.46:1 in weight). The scale bar denotes 50 μm. (Reprinted with permission from [118]. Copyright 2017 American Chemical Society.)

The interest in such a stable cholesteric[1] liquid-crystalline phase was further exploited by several groups [36, 119–121] for the obtaining of cellulose-polymer iridescent[2] films. Ionic and nonfunctional polymers/monomers were used (Table 8.2).

The self-assembly of the CNCs into cholesteric phases is ascribed to their twisted (left-handed helicoidal) shape and anisotropic charge distribution [125] and the formation of ordered phases requires neutral, fully water-soluble polymers. Thus, the electrostatically driven self-assembly of the colloidal-scale CNCs will not be altered by changes in the CNC charge distribution or excessive CNC agglomeration [36, 119]. In a very interesting biomimicking approach, Malho et al. [122, 123] have designed cellulose binding proteins with a natural tendency toward multimer complex formation as an adhesive matrix for combinations with

1 Chiral nematic
2 Changing color with illumination or observation angle

nanofibrillated cellulose. Their findings show that the protein matrix affects the material mechanic properties mainly through interactions during plastic deformation and that the dynamic rearrangements lead to increased interactions between fibrils over higher length scales. Other studies, scarcely found in the literature, reveal that supramolecular grafting can also be used in the design of light-healable supramolecular nanocomposites, with significantly improved mechanical properties [47]. In this respect, Biyani et al. [124] and McKee et al. [126] developed nanocomposites based on a telechelic poly(ethylene-*co*-butylene) functionalized with ureidopyrimidone (UPy) and CNCs decorated with the same binding motif. Again, the nanocomposites show better mechanical properties. Moreover, when exposed to ultraviolet radiation, deliberately introduced defects are healed quickly and efficiently. This is because light-excited UPy motifs convert the absorbed energy into heat, thus causing temporary disengagement of the hydrogen-bonding motifs and concomitant reversible decrease of the supramolecular polymer molecular weight and viscosity. The results are valid even at a filler content of 20% w/w, that is, in compositions that exhibit high strength and stiffness.

Table 8.2 Nanocellulose use for (1) crustacean-like cholesteric iridescent films, (2) healable nanocomposites, and (3) adhesive matrices

Nanocellulose	Polymer/monomer	Application	Ref.
Microfibrils	PCL	1	[117]
	Protein	3	[122, 123]
Nanocrystals	PHEMA	1	[118]
	Poly(ethylene glycol)	1	[119]
	Anionic sodium poly(acrylate)	1	[119]
	Urea Formaldehyde	1	[120]
	Poly(vinyl alcohol)	1	[36]
	Poly(oligoethylene glycol methacrylate-*co*-2-ureido-4-pyrimidone methacrylate)	1	[121]
	Telechelic poly(ethylene-*co*-butylene) functionalized with ureidopyrimidone (UPy)	2	[124]

Further studies are needed for a more precise understanding of the mechanisms behind this passionate supramolecular grafting of nanocellulose.

8.3.5.2 Starch nanoparticles

Insofar as SNPs are recovered and processed, nanocrystals are not only poorly dispersible in solvents generally used with polymers (due to reaggregation via strong hydrogen bonding), but also characterized by a very low thermal stability [7, 93, 127]. A negative SNP melting, specifically with water trace, has to be avoided during processing. In order to avoid it, the solvent casting method is often used to prepare SNPs/polymer composites. The interest in starch nanocrystals is related to their platelet-like structures suitable for possible improvement of barrier properties. However, to stimulate industrial-scale use, melt blend preparation has to be investigated [57]. For that, the surface modification of SNPs is required [127–130] to confer customized functions to expand the SNP applications. As CNCs and ChNCs, reactive hydroxyl groups are present on starch nanocrystals surfaces providing modifications by appropriate chemical reactions. However, the investigations of composite materials incorporating these particles are again relatively limited, that is, poly(styrene-*co*-butyl acrylate), natural rubber, pullulan, thermoplastic starch, polyvinyl alcohol, soy protein, or waterborne polyurethane [92] have been mixed up with SNPs. Four strategies have already been investigated in the literature [57]: (i) modification by chemical modification with small molecules, (ii) *grafting from* polymer chains with polymerization of a monomer (Fig. 8.1a), (iii) *grafting onto* polymer chains with coupling agents (Fig. 8.1b), and (iv) *supramolecular grafting* (Fig. 8.1d).

The feasibility of SNPs surface modification was first confirmed by Angellier et al. [95]. In this study, after sulfuric acid hydrolysis of native starch granules, SNPs obtained were superficially modified using two different reagents, that is, alkenyl succinic anhydride (ASA) and phenyl isocyanate (PI). After surface chemical treatment, the platelet-like geometric form of the PI-modified starch nanocrystals seemed preserved even though the size of the nano-objects was decreased (Fig. 8.12) [111]. The PI-modified starch nanocrystals dispersed well in methylene chloride solution, contrary to unmodified SNPs. Therefore, isocyanate functions modify the polarity of nanostarch allowing to melt process composite materials using nonpolar polymers as matrices [111].

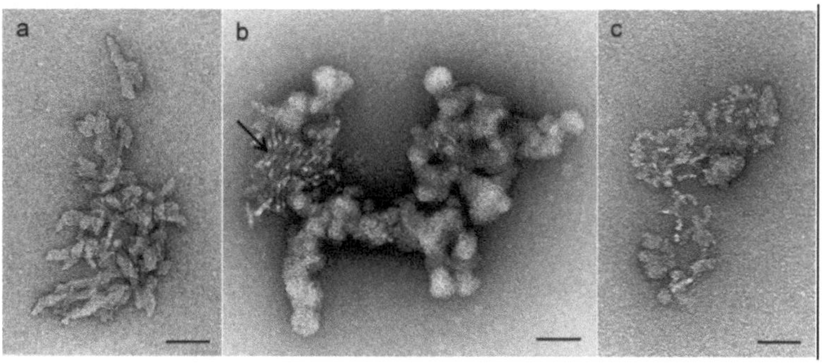

Figure 8.12 TEM micrographs of waxy maize starch nanocrystals (a) before and after chemical treatment with (b) alkenyl succinic anhydride (ASA) and (c) phenyl isocyanate (PI). Scale bar: 50 nm. Reprinted with permission from [111]. Copyright (2019) American Chemical Society.

The modification of SNPs with acetic anhydride also leads to promising results [57, 131]. With hydrophobic property and improved solubility in common organic solvents, the crystalline structure of acetylated starch nanocrystals was also changed from A-style to V-style. In addition, the platelet-like starch nanocrystals became sphere shaped after modification and the size increased from the original 20–40 nm to 63–271 nm. [57, 131]. However, surface chemical modification with small molecule attachment does not always provide miscibility between nanoparticles and polymers. In this regard, grafting methods are as usual exploited.

8.3.5.2.1 Grafting onto SNPs

Grafting onto SNPs was investigated with various preformed polymers using different coupling agents. The studies are exhaustively discussed by Lin *et al.* [57], concluding that SNPs surface modification needs mild conditions (temperature, pH) to preserve morphology integrity. Table 8.3 summarizes the conditions for the main *grafting onto* SNPs modifications including native starch source. In this case also, *grafting onto* suffers from drawbacks, that is, low reaction control and weak grafting, mainly to long-chain polymers. In a manner to overcome these drawbacks, the *grafting from* strategy was also investigated with SNPs.

Table 8.3 Surface modification of nanostarch via grafting

Source	Grafting technique: Method	Grafting technique: Variation	Modification	Reference
Waxy Maize	Chemical reaction	–	PI	[111]
Corn	Chemical reaction	–	AA	[131]
Waxy Maize	Onto	2,4-TDI surface modification	PEGME	[130]
Waxy Maize	Onto	2,4-TDI surface modification	PTHF, PPGBE, PCL	[134]
Potato and Waxy Maize	Onto	Esterification	Aliphatic chloride	[135]
Pea	From	ROP	PCL	[132, 136]
Potato	From	ROP	PCL	[129]
Corn	From	FRP	PS	[133]
Waxy Maize	Supramolecular	ROP; ATRP	PDLA-*b*-PGMA	[137]

TDI: Toluene diisocyanate; **PI**: Phenyl isocyanate; **AA**: Acetic anhydride; **PEGME**: Poly(ethylene glycol) methyl ether; PPGBE: Poly(propylene glycol) monobutyl ether; **PTHF**: Poly(tetrahydrofuran); **PCL**: Poly(caprolactone); **PS**: Polystyrene; **PDLA-*co*-PGMA**: Poly(D-lactide)-*co*-poly(glycidyl methacrylate)

8.3.5.2.2 Grafting from SNPs

Both microwave-assisted and thermal ROP of caprolactone to PCL [128, 129, 132] have been studied, as well as the free radical polymerization (FRP) for polystyrene (PS) [133]. With *grafting from* strategy, a higher grafting density can be realized, and the properties of starch nanocrystals can be regulated through the selection and control of the grafting polymer chains length and type. The grafted SNPs were found to have improved compatibility with a PLA matrix. Grafting PCL polymer chains to starch nanocrystals using bulk polymerization was also successfully used [129] leading to crystalline structure and morphology of nanocrystals unaltered.

As an alternative to ROP, starch nanocrystals were modified by surface induced free radical polymerization of styrene. A starch nanocrystal copolymer was prepared by graft copolymerization of starch nanocrystals with styrene [133]. After grafting the hydrophobic polystyrene chains, the modified starch nanocrystals became amphiphilic nanoparticles with SNP sizes increased from 50 nm nanoplatelet-like morphology to 80–100 nm spherical morphology. Table 8.3 summarizes the surface modification of nanostarch via grafting.

Some issues, however, are to be discussed in terms of morphology preservation during grafting and/or melt processing. In this respect, as a method proceeding in milder conditions, *supramolecular grafting* can be used.

8.3.5.2.3 Supramolecular grafting on SNPs

Fully biobased and biodegradable nanocomposites based on poly(L-lactide) (PLLA) and starch nanoplatelets (SNPs) were prepared by Benali et al. [137] using an original strategy involving supramolecular chemistry. To this end, poly(D-lactide)-*b*-poly(glycidyl methacrylate) (PDLA-*b*-PGMA) was first synthesized via the combination of ring-opening polymerization (ROP) and atom transfer radical polymerization (ATRP). NMR spectroscopy and SEC analysis confirmed an efficient control over the copolymer synthesis. The SNPs were then blended with the copolymer for producing a PDLA-*b*-PGMA/SNPs masterbatch. The solvent casting method was studied to improve the SNPs thermal resistance and their compatibility with the PLLA matrix. A masterbatch (PDLA-*b*-PGMA/SNPs) was obtained by solvent casting with specific attention to the solvent selection to preserve SNPs morphology. The copolymer-SNPs supramolecular interactions taking place with hydrogen bonding are highlighted using near-infrared (NIR) spectroscopy. Thereafter, the masterbatch was melt-blended with virgin PLLA and then a thin film of PLLA/PDLA-*b*-PGMA/SNPs nanocomposites was melt-processed by compression molding. The obtained nanocomposites films were by thermogravimetric analysis (TGA) and differential scanning calorimetry (DSC). Our findings allow us to conclude that supramolecular interactions, that is, stereocomplexation between the PLLA matrix and the PDLA block of the copolymer formed on one side and hydrogen bonding between SNPs and the PGMA block of the copolymer on the other side led to a synergetic effect with the maintenance of SNPs nanoplatelets and their morphology during melt processing. Quartz crystal microbalance (QCM) highlighted a promising effect on PLLA barrier properties against water vapor.

8.3.5.3 Chitin nanoparticles

Chitin nanocrystals (extraction presented in Fig. 8.13 [138]) attract attention with their unique cationic structure [139]. Recently, they were used as reinforcing agents in different polymeric matrices, such as natural rubber [140] and polycaprolactone [141]. However, as with all other hydrophilic nanoparticles, nanochitin self-aggregates easily and is immiscible with hydrophobic polymer matrices. Several studies deal with surface modification of chitin nanoparticles to introduce small lipophilic molecules such as stearic acid chloride, alkenyl succinic anhydride, and phenyl isocyanate. They are summarized and well discussed in the review of Dufresne from 2010 [142] and will not be discussed here.

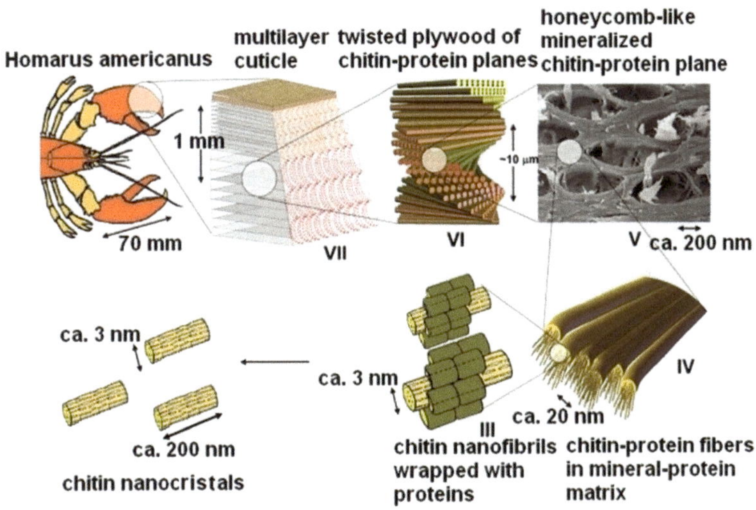

Figure 8.13 Schematic representation of chitin nanocrystals extraction. Adapted with permission from [138].

Besides, a number of scientific papers report on the surface modification of chitin nanoparticles via grafting of polymer chains (Table 8.4). *Grafting from* and *grafting onto* were successfully applied to surface modify chitin nanofibers and nanocrystals.

Table 8.4 Surface modification of nanochitin via grafting

		Grafting technique			
Nanochitin	**Source**	**Method**	**Variation**	**Polymer**	**Ref.**
Nanofibers	Crab shells	From	ROP	Poly(lactide-*co*-caprolactone)	[143]
	Unknown	From	FRP	Poly(acrylic acid)	[144]
Nanocrystals (whiskers)	Crab shells	From	ROP	Poly(caprolactone)	[141]
	Unknown	From	ROP	Poly(lactide)	[145, 146]
	Unknown	Onto	Esterification	Methoxy poly(ethylene glycol) acid	[147]
	Shrimp shells	Onto	Thionyl chloride activated esterification	Poly(hydroxybutyrate-*co*-hydroxyvalerate)	[148]

8.3.5.3.1 Grafting from nanochitin

Most of the related studies here take advantage of the fast and relatively easy *grafting from* technique (no steric hindrance and low viscosity of the reaction medium) for introducing hydrophilic and hydrophobic polymer chains. For example, Ifuku et al. [144] use the free-radical polymerization of acrylic acid (AA) with persulfate as an initiator in an aqueous medium to introduce poly(acrylic acid) (PAA) chains on the surface of chitin nanofibers (Fig. 8.14). Multiple analytical methods showed AA was grafted on the surface and in the amorphous part, thus preserving the original crystal structure of the chitin – preserved chitin nanofibers morphology after polymerization and with efficient dissociation and homogeneous dispersion due to electrostatic repulsion between the PAA-grafted nanofibers.

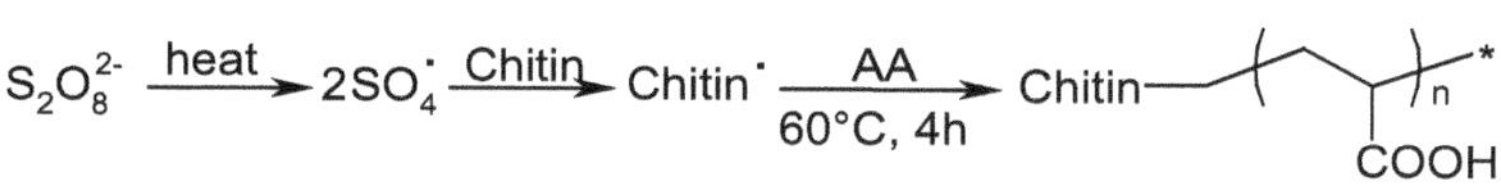

Figure 8.14 Free-radical graft polymerization of AA on chitin nanofibers.

Other groups tried ROP for polyester grafting on chitin nanofibers and whiskers [141, 143, 145, 146]. Thus, poly(lactide-

co-caprolactone) (P(LA-*co*-CL)), PCL, and PLA were successfully *introduced* in order to disperse nanochitin into a PLLA matrix [145, 146] or to directly form nanocomposites [141, 143]. In all cases, the initial structure of nanochitin was remarkably preserved and the tensile strength and elongation at break as well as the hydrophobicity of the nanocomposites were significantly improved.

However, the possible residues of non-grafted homopolymers as well as the undefined polymer chain characteristics still presented a disadvantage of the method. Therefore, the groups of Wang et al. [148] and Mol et al. [147] applied the *grafting onto* approach.

8.3.5.3.2 *Grafting onto nanochitin*

So far, there are only two studies dealing with the *grafting onto* nanochitin [147, 148], both using chitin whiskers as the starting material. The group of Wang et al. [148] obtained chemically modified chitin nanocrystals by grafting poly(3-hydroxybutyrate-*co*-3-hydroxyvalerate) (PHBV) onto chitin backbone via chlorination while preserving the amino group. Analyses revealed successful grafting and preserved whiskers structure but with a modified appearance. A large amount of white dots (suggested to be PHBV) surrounded the chitin and blurred the outlines of the nanocrystals, while the degree of aggregation seemed to be reduced [148]. As expected, contact angle measurement showed that improved hydrophobicity of chitin whiskers and also found to suppress PHBV crystallization.

In an attempt to enhance recyclability of acrylonitrile-butadiene-styrene (ABS) rubber, Mol et al. [147] derived chitin whiskers surface grafted with methoxy poly(ethylene glycol) (mPEG) of different molar masses. Indeed, nowadays recyclability of polymeric materials is a very important question and gathers growing attention from both universities and industry. The major problem with this reprocessing comes from the severe damage to the molecular architecture and microstructure of the polymer, which often results in poorer mechanical properties of the recycled material. A possible solution considers the incorporation of nanoparticles as reinforcing agents. In this case [147], the modified chitin whiskers were incorporated into reprocessed ABS (acrylonitrile–butadiene–styrene) to yield nanocomposites with 0.5% (mass/mass) whiskers. The results showed that high molar mass mPEG grafts increase the strength, elongation at break, and stiffness of the reprocessed ABS over virgin, reprocessed ABS and reprocessed ABS/unmodified whiskers. This indicates that the use of surface-modified chitin whiskers can be valuable in improving the mechanical properties of recycled polymers and, consequently, enhancing their recyclability.

8.4 Conclusions

The current progress in surface modification of biobased polysaccharide nanoparticles, for example, nanocelluloses, starch, and chitin nanoparticles via grafting techniques, was carefully examined. Description about the different methods, for example, *grafting from*, *grafting onto*, *grafting through*, and *supramolecular grafting*, is provided in the first section. Further, the current state of the art in polysaccharide particle preparation and physical and chemical properties was renewed, including very recent trends and strategies. The final section thoroughly discussed the latest methods and achievements toward surface-modified nanocelluloses, nanochitin, and nanostarch particles via *grafting from*, *grafting onto* and most importantly *supramolecular grafting*. Thus, this chapter summarizes the most recent trends in the field of functional nanomaterials from bioderived polysaccharide nanoparticles.

Abbreviations

AA	Acrylic acid
AcA	Acetic anhydride
AD	Adamantane
ASA	Alkenyl succinic anhydride
ATRP	Atom transfer radical polymerization
CD	Cyclodextrins
ChNCs	Chitin nanocrystals
ChNFs	Chitin nanofibers
CNCs	Cellulose nanocrystals
CNFs	Cellulose nanofibrils
DMAC	*N,N*-Dimethylacetamide
DMF	Dimethylformamide
DSC	Differential scanning calorimetry

FRP	Free radical polymerization
FTIR-ATR	Fourier-transform infrared attenuated total reflectance spectroscopy
HEMA	2-Hydroxyethyl methacrylate
HFIP	1,1,1,3,3,3-Hexafluoro-2-propanol
k_a	Activation constant
k_{add}	Addition constant
k_{-add}	Dissociation constant
k_d	Deactivation constant
LA	Lactide
LiCl	Lithium chloride
Me^z-Y/L	Transition metal complex (activator, where Y may be another differ from the ligand (L) or be a counter ion)
NIR	Near-infrared spectroscopy
NMP	Nitroxide-mediated polymerization
NMR	Nuclear magnetic resonance spectroscopy
NO*	Stable nitroxide radical
PAA	Poly(acrylic acid)
PCL	Poly(ε-caprolactone)
PDLA	Poly(D-lactide)
PDLA-*co*-PGMA	Poly(D-lactide)-*co*-poly(glycidyl methacrylate)
PEGME	Poly(ethylene glycol) methyl ether
PHEMA	Poly(2-hydroxyethyl methacrylate)
PI	Phenyl isocyanate
PLA	Polylactide
PLLA	Poly(L-lactide)
Pn*	Active propagating species
Pn-X	Dormant alkyl halide species
Pn–X	Dormant species
PPGBE	Poly(propylene glycol) monobutyl ether
PS	Polystyrene
PSNC	Polysaccharide nanocrystal
PTHF	Poly(tetrahydrofuran)
QCM	Quartz crystal microbalance
RAFT	Reversible addition–fragmentation chain transfer polymerization
RDRP	Reversible-deactivation radical polymerization
SEC	Size-exclusion chromatography

SEM	Scanning electron microscopy
SI	Surface initiated
SI-CRP	Surface-initiated controlled radical polymerization
SI-FRP	Surface-initiated free radical polymerization
SI-ROP	Surface-initiated ring-opening polymerization
$Sn(Oct)_2$	Tin(II) 2-ethylhexanoate
SNPs	Starch nanocrystals
TAPPI	The Technical Association of the Pulp and Paper Industry
TDI	Toluene diisocyanate
TEMPO	2,2,6,6-Tetramethylpiperidine-1-oxyl
TGA	Thermogravimetric analysis
UPy	Ureidopyrimidone
$X\cdots Me^{z+1}$–Y/L	Metal halide complex
XPS	X-ray photoelectron spectroscopy
ε-CL	ε-Caprolactone

References

1. R. Mülhaupt, Hermann Staudinger and the origin of macromolecular chemistry, *Angew. Chemie Int. Ed.* 43 (2004) 1054–1063. doi:10.1002/anie.200330070.
2. S.K. Kumar, B.C. Benicewicz, R.A. Vaia, K.I. Winey, *50th Anniversary Perspective*: Are Polymer Nanocomposites Practical for Applications? *Macromolecules* (2017) acs.macromol.6b02330. doi:10.1021/acs.macromol.6b02330.
3. Y. Zheng, J. Monty, R.J. Linhardt, Polysaccharide-based nanocomposites and their applications, *Carbohydr. Res.* 405 (2015) 23–32. doi:10.1016/j.carres.2014.07.016.
4. N. Lin, J. Huang, A. Dufresne, Preparation, properties and applications of polysaccharide nanocrystals in advanced functional nanomaterials: A review, *Nanoscale* (2012).
5. N. Lin, G. Chen, J. Huang, A. Dufresne, P.R. Chang, Effects of polymer-grafted natural nanocrystals on the structure and mechanical properties of poly(lactic acid): A case of cellulose whisker-graft-polycaprolactone, *J. Appl. Polym. Sci.* 113 (2009) 3417–3425. doi:10.1002/app.30308.
6. M.V. Tzoumaki, T. Moschakis, V. Kiosseoglou, C.G. Biliaderis, Oil-in-water emulsions stabilized by chitin nanocrystal particles, *Food Hydrocoll.* 25 (2011) 1521–1529. doi:10.1016/j.foodhyd.2011.02.008.

7. D. Le Corre, J. Bras, A. Dufresne, Starch nanoparticles: A review, *Biomacromolecules* 11 (2010) 1139–53. doi:10.1021/bm901428y.
8. D. LeCorre, J. Bras, A. Dufresne, Influence of native starch's properties on starch nanocrystals thermal properties, *Carbohydr. Polym.* 87 (2012) 658–666. doi:10.1016/j.carbpol.2011.08.042.
9. S.J. Eichhorn, C.A. Baillie, N. Zafeiropoulos, L.Y. Mwaikambo, M.P. Ansell, A. Dufresne, et al., Review: Current international research into cellulosic fibres and composites, *J. Mater. Sci.* 36 (2001) 2107–2131. doi:10.1023/a:1017512029696.
10. P. Gatenholm, D. Klemm, Bacterial nanocellulose as a renewable material for biomedical applications, *MRS Bull.* 35 (2011) 208–213. doi:10.1557/mrs2010.653.
11. M. Valodkar, S. Thakore, Organically modified nanosized starch derivatives as excellent reinforcing agents for bionanocomposites, *Carbohydr. Polym.* 86 (2011) 1244–1251. doi:http://dx.doi.org/10.1016/j.carbpol.2011.06.020.
12. Y. Habibi, Key advances in the chemical modification of nanocelluloses, *Chem. Soc. Rev.* 43 (2014) 1519–1542. doi:10.1039/c3cs60204d.
13. M. Börjesson, G. Westman, Crystalline nanocellulose: preparation, modification, and properties, in: M. Poletto, H.L. Omaghi (Eds.), *Cellulose - Fundamental Aspects and Current Trends,* InTech, 2015: p. 282.
14. MarketResearch.com, The Global Market for Nanocellulose: Opportunity, Markets, Applications, Technologies and Companies, 2016.
15. H.P.S. Abdul Khalil, Y. Davoudpour, C.K. Saurabh, M.S. Hossain, A.S. Adnan, R. Dungani, et al., A review on nanocellulosic fibres as new material for sustainable packaging: Process and applications, *Renew. Sustain. Energy Rev.* 64 (2016) 823–836. doi:10.1016/j.rser.2016.06.072.
16. M.S. Reid, M. Villalobos, E.D. Cranston, Benchmarking cellulose nanocrystals : From the laboratory to industrial production, *Langmuir* 33 (2017) 1583–1598.
17. Y. Habibi, A. Dufresne, Highly filled bionanocomposites from functionalized polysaccharide nanocrystals, *Biomacromolecules* 9 (2008) 1974–1980. doi:10.1021/bm8001717.
18. D. Roy, M. Semsarilar, J.T. Guthrie, S. Perrier, Cellulose modification by polymer grafting: A review, *Chem. Soc. Rev.* 38 (2009) 2046–2064. http://dx.doi.org/10.1039/B808639G.

19. K. Schneider, Mechanical properties of polymers at surfaces and interfaces, *Polym. Surf. Interfaces* (2008) 139–160. doi:10.1007/978-3-540-73864-0.
20. H. Kang, R. Liu, Y. Huang, Graft modification of cellulose: Methods, properties and applications, *Polymer (United Kingdom)* 70 (2015) A1–A16. doi:10.1016/j.polymer.2015.05.041.
21. G. Moad, D.H. Solomon, *The Chemistry of Radical Polymerization*, Elsevier B.V., 2006.
22. F.L. Hatton, S.A. Kedzior, E.D. Cranston, A. Carlmark, Grafting-from cellulose nanocrystals via photoinduced Cu-mediated reversible-deactivation radical polymerization, *Carbohydr. Polym.* 157 (2016). doi:10.1016/j.carbpol.2016.10.064.
23. M.N. Gandini, Alessandro Belgacem, The surface and in-depth modification of cellulose fibers, *Adv. Polym. Sci.* 271 (2016) 169–206.
24. W.H. Daly, T.S. Evenson, S.T. Iacono, R. Walker Jones, Recent developments in cellulose grafting chemistry utilizing Barton ester intermediates and nitroxide mediation, *Macromol. Symp.* 174 (2001) 155–163. doi:10.1002/1521-3900(200109)174:1<155::AID-MASY155>3.0.CO;2-O.
25. J. Yi, Q. Xu, X. Zhang, H. Zhang, Chiral-nematic self-ordering of rodlike cellulose nanocrystals grafted with poly(styrene) in both thermotropic and lyotropic states, *Polymer (Guildf)* 49 (2008) 4406–4412. doi:10.1016/j.polymer.2008.08.008.
26. J. Yi, Q. Xu, X. Zhang, H. Zhang, Temperature-induced chiral nematic phase changes of suspensions of poly(N,N-dimethylaminoethyl methacrylate)-grafted cellulose nanocrystals, *Cellulose* 16 (2009) 989–997. doi:10.1007/s10570-009-9350-9.
27. G. Morandi, L. Heath, W. Thielemans, Cellulose nanocrystals grafted with polystyrene chains through surface-initiated atom transfer radical polymerization (SI-ATRP), *Langmuir* 25 (2009) 8280–8286. doi:10.1021/la900452a.
28. E. Zeinali, V. Haddadi-Asl, H. Roghani-Mamaqani, Nanocrystalline cellulose grafted random copolymers of N-isopropylacrylamide and acrylic acid synthesized by RAFT polymerization: Effect of different acrylic acid contents on LCST behavior, *RSC Adv.* 4 (2014) 31428–31442. doi:10.1039/C4RA05442C.
29. E. Malmstrom, A. Carlmark, Controlled grafting of cellulose fibres – An outlook beyond paper and cardboard, *Polym. Chem.* 3 (2012) 1702–1713. http://dx.doi.org/10.1039/C1PY00445J.

30. G. Odian, *Principles of Polymerization*, IVth, John Wiley & Sons, Hoboken, 2004.

31. K. Matyjaszewski, T. P. Davis (Eds.), *Handbook of Radical Polymerization*, John Wiley & Sons, Hoboken, 2002: pp. 361–406.

32. C. Jérôme, P. Lecomte, Recent advances in the synthesis of aliphatic polyesters by ring-opening polymerization, *Adv. Drug Deliv. Rev.* 60 (2008) 1056–1076. doi:10.1016/j.addr.2008.02.008.

33. S. Hansson, V. Trouillet, T. Tischer, A.S. Goldmann, A. Carlmark, C. Barner-Kowollik, et al., Grafting efficiency of synthetic polymers onto biomaterials: A comparative study of grafting-from versus grafting-to, *Biomacromolecules.* 14 (2013) 64–74. doi:10.1021/bm3013132.

34. S. Yoshikawa, T. Takayama, N. Tsubokawa, Grafting reaction of living polymer cations with amino groups on chitosan powder, *J. Appl. Polym. Sci.* 68 (1998) 1883–1889. doi:10.1002/(SICI)1097-4628(19980613)68:11<1883::AID-APP21>3.0.CO;2-U.

35. N. Tsubokawa, T. Iida, T. Takayama, Modification of cellulose powder surface by grafting of polymers with controlled molecular weight and narrow molecular weight distribution, *J. Appl. Polym. Sci.* 75 (2000) 515–522. doi:10.1002/(sici)1097-4628(20000124)75:4<515::aid-app6>3.0.co;2-q.

36. H. Wang, J. He, M. Zhang, K.C. Tam, P. Ni, A new pathway towards polymer modified cellulose nanocrystals via a "grafting onto" process for drug delivery, *Polym. Chem.* 6 (2015) 4206–4209. doi:10.1039/c5py00466g.

37. J. Chen, N. Lin, J. Huang, A. Dufresne, Highly alkynyl-functionalization of cellulose nanocrystals and advanced nanocomposites thereof via click chemistry, *Polym. Chem.* 6 (2015) 4385–4395. doi:10.1039/C5PY00367A.

38. A. Benkaddour, K. Jradi, S. Robert, C. Daneault, Grafting of polycaprolactone on oxidized nanocelluloses by click chemistry, *Nanomaterials.* 3 (2013) 141–157. doi:10.3390/nano3010141.

39. A.P. Mangalam, J. Simonsen, A.S. Benight, Cellulose/DNA hybrid nanomaterials, Biomacromolecules (2009) 497–504. doi:10.1021/bm800925x.

40. N. Ljungberg, C. Bonini, F. Bortolussi, C. Boisson, L. Heux, J.Y. Cavaillé, New nanocomposite materials reinforced with cellulose whiskers in atactic polypropylene: Effect of surface and dispersion characteristics, *Biomacromolecules* 6 (2005) 2732–2739. doi:10.1021/bm050222v.

41. F. Azzam, L. Heux, B. Jean, J.-L. Putaux, Preparation by grafting onto, characterization and properties of thermally responsive polymer-

decorated cellulose nanocrystals, *Biomacromolecules* 11 (2010) 3652–3659.

42. S. Harrisson, G.L. Drisko, E. Malmström, A. Hult, K.L. Wooley, Hybrid rigid/soft and biologic/synthetic materials: Polymers grafted onto cellulose microcrystals, *Biomacromolecules* 12 (2011) 1214–1223. doi:10.1021/bm101506j.
43. J. Araki, M. Wada, S. Kuga, Steric stabilization of a cellulose microcrystal suspension by poly(ethylene glycol) grafting, *Cellulose* (2001) 21–27. doi:10.1021/la001070m.
44. A. Pei, J.M. Malho, J. Ruokolainen, Q. Zhou, L.A. Berglund, Strong nanocomposite reinforcement effects in polyurethane elastomer with low volume fraction of cellulose nanocrystals, *Macromolecules* 44 (2011) 4422–4427. doi:10.1021/ma200318k.
45. X. Cao, Y. Habibi, L. A. Lucia, One-pot polymerization, surface grafting, and processing of waterborne polyurethane-cellulose nanocrystal nanocomposites, *J. Mater. Chem.* 19 (2009) 7137–7145. doi:10.1039/b910517d.
46. E. Kloser, D.G. Gray, Surface grafting of cellulose nanocrystals with poly(ethylene oxide) in aqueous media, *Langmuir* 26 (2010) 13450–13456. doi:10.1021/la101795s.
47. Z. Li, X. Zhou, C. Pei, Synthesis and characterization of MPS-g-PLA copolymer and its application in surface modification of bacterial cellulose, *Int. J. Polym. Anal. Charact.* 15 (2010) 199–209. doi:10.1080/10236661003681222.
48. R. Mincheva, L. Jasmani, T. Josse, Y. Paint, J.M. Raquez, P. Gerbaux, et al., Binary mixed homopolymer brushes tethered to cellulose nanocrystals: A step towards compatibilized polyester blends, *Biomacromolecules* 17 (2016) 3048–3059. doi:10.1021/acs.biomac.6b00932.
49. V. K. Thakur, *Cellulose-Based Graft Copolymers: Structure and Chemistry*, Taylor & Francis Group, Washington, D.C., 2015. doi:https://doi.org/10.1201/b18390.
50. A. Bhattacharya, B.N. Misra, Grafting: A versatile means to modify polymers: Techniques, factors and applications, *Prog. Polym. Sci.* 29 (2004) 767–814. doi:10.1016/j.progpolymsci.2004.05.002.
51. G. V. Oshovsky, D.N. Reinhoudt, W. Verboom, Supramolecular chemistry in water, *Angew. Chemie Int. Ed.* 46 (2007) 2366–2393. doi:10.1002/anie.200602815.
52. R. Sengupta, S. Chakraborty, S. Bandyopadhyay, S. Dasgupta, R. Mukhopadhyay, K. Auddy, et al., A short review on rubber/clay

nanocomposites with emphasis on mechanical properties, *Polymer Engineering and Science.* 47 (2007) 21–25. doi:10.1002/pen.

53. C.J. Chirayil, L. Mathew, S. Thomas, Review of recent research in nano cellulose preparation from different lignocellulosic fibers, *Rev. Adv. Mater. Sci.* 37 (2014) 20–28.

54. I.A. Sacui, R.C. Nieuwendaal, D.J. Burnett, S.J. Stranick, M. Jorfi, C. Weder, et al., Comparison of the properties of cellulose nanocrystals and cellulose nanofibrils isolated from bacteria, tunicate, and wood processed using acid, enzymatic, mechanical, and oxidative methods, *ACS Appl. Mater. Interfaces* 6 (2014) 6127–6138. doi:10.1021/am500359f.

55. J. Kadokawa, Fabrication of nanostructured and microstructured chitin materials through gelation with suitable dispersion media, *RSC Adv.* 5 (2015) 12736–12746. doi:10.1039/C4RA15319G.

56. A.M. Shi, D. Li, L.J. Wang, B.Z. Li, B. Adhikari, Preparation of starch-based nanoparticles through high-pressure homogenization and miniemulsion cross-linking: Influence of various process parameters on particle size and stability, *Carbohydr. Polym.* (2011) 1604–1610. doi:10.1016/j.carbpol.2010.10.011.

57. N. Lin, J. Huang, P.R. Chang, D.P. Anderson, J. Yu, Preparation, modification, and application of starch nanocrystals in nanomaterials: a review, *J. Nanomater.* 2011 (2011) 20. doi:10.1155/2011/573687.

58. Y. Qin, C. Liu, S. Jiang, L. Xiong, Q. Sun, Characterization of starch nanoparticles prepared by nanoprecipitation: Influence of amylose content and starch type, *Ind. Crops Prod.* 87 (2016) 182–190. doi:10.1016/j.indcrop.2016.04.038.

59. J. Huang, P.R. Chang, A. Dufresne, Polysaccharide nanocrystals: current status and prospects in material science in: J. Huang, P.R. Chang, A. Dufresne (Eds.), *Polysaccharide-Based Nanocrystals: Chemistry and Applications,* Wiley-VCH Verlag GmbH & Co. KGaA, Weinheim, Germany, 2015: pp. 1–14.

60. R.F. Nickerson, J.A. Habrle, Cellulose intercrystalline, *Ind. Engeenering Chem.* 39 (1947) 1507–1512.

61. Roadmap for the Development of International Standards for Nanocellulose, Arlington, Virginia, USA, 2011. Available at: https://www.tappinano.org/media/1468/2011-roadmapfornanocellulosestandards.pdf.

62. G. Siqueira, J. Bras, A. Dufresne, Cellulosic bionanocomposites: A review of preparation, properties and applications, *Polymers (Basel)* 2 (2010) 728–765. doi:10.3390/polym2040728.

63. R.J. Moon, A. Martini, J. Nairn, J. Simonsen, J. Youngblood, Cellulose nanomaterials review: Structure, properties and nanocomposites, *Chem. Soc. Rev.* 40 (2011) 3941–3994. doi:10.1039/c0cs00108b.

64. J. Huang, P.R. Chang, N. Lin, A. Dufresne, *Polysaccharide-Based Nanocrystals: Chemistry and Applications*, Chemical I, Wiley-VCH Verlag GmbH & Co. KGaA, Weinheim, Germany, 2015.

65. Y. Habibi, L.A. Lucia, O.J. Rojas, Cellulose nanocrystals: Chemistry, self-assembly, and applications, *Chem. Rev.* (Washington, DC, United States) 110 (2010) 3479–3500. doi:10.1021/cr900339w.

66. S.M. Mukherjee, H.J. Woods, X-ray and electron microscope studies of the degradation of cellulose by sulfuric acid, *Biochim. Biophys. Acta.* 10 (1953) 499–511.

67. B. Wang, M. Sain, K. Oksman, Study of structural morphology of hemp fiber from the micro to the nanoscale, *Appl. Compos. Mater.* 14 (2007) 89–103.

68. J.N. Saddler, H.H. Brownell, L.P. Clermont, N. Levitin, Enzymatic hydrolysis of cellulose and various pretreated wood fractions, *Biotechnol. Bioeng.* 24 (1982) 1389–1402.

69. A.M.D. Lima, J.T. Wong, M. Paillet, R. Borsali, R. Pecora, Translational and rotational dynamics of rodlike cellulose whiskers, *Langmuir* 19 (2003) 24–29.

70. M.M. Lima, R. Borsali, Static and dynamic light scattering from polyelectrolyte microcrystal cellulose, *Langmuir* 18 (2002) 992–996.

71. P.L. Beltrame, E.D. Paglia, A. Seves, E. Pellizzoni, M. Romano, Structural features of native cellulose gels and films from their susceptibility to enzymatic attack, *J. Appl. Polym. Sci.* 44 (1992) 2095–2101.

72. H.R. Hohl, J. Jehli, Presence of cellulose microfibrils in proteinaceous slime track of *Dictyostelium discoideum, Archiv für Mikrobiologie,* 92 (1973) 179–187.

73. J.A. Kelly, A.M. Shukaliak, C.C.Y. Cheung, K.E. Shopsowitz, W.Y. Hamad, M.J. MacLachlan, Responsive photonic hydrogels based on nanocrystalline cellulose, *Angew. Chem. Int. Ed.* 52 (2013) 8912–8916.

74. H. Fukuzumi, T. Saito, T. Wata, Y. Kumamoto, A. Isogai, Transparent and high gas barrier films of cellulose nanofibers prepared by TEMPO-mediated oxidation, *Biomacromolecules* 10 (2009) 162–165.

75. J. Mendes, P.K. Annamalai, S.J. Eichhorn, R. Rusli, S.J. Rowan, E.J. Foster, et al., Bioinspired mechanically adaptive polymer nanocomposites with water-activated shape-memory effect, *Macromolecules* 44 (2011) 6827–6835.

76. M.V. Biyani, E.J. Foster, C. Weder, Light-healable supramolecular nanocomposites based on modified cellulose nanocrystals, *ACS Macro Lett.* 2 (2013) 236–240.

77. K.A. Potter, M. Jorfi, K.T. Householder, E.J. Foster, C. Weder, J.R. Capadona, Curcumin-releasing mechanically adaptive intracortical implants improve the proximal neuronal density and blood-brain barrier stability, *Acta Biomater.* 10 (2014) 2209–2222.

78. J.R. Capadona, K. Shanmuganathan, D.J. Tyler, S.J. Rowan, C. Weder, Stimuli-responsive polymer nanocomposites inspired by the sea cucumber dermis, *Science* 319 (2008) 1370–1374.

79. Rojas, O. J. *Cellulose Chemistry and Properties: Fibers, Nanocelluloses and Advanced Materials.* Part of the series *Advances in Polymer Science,* Vol. 271, Springers, 2016.

80. P.-L. Bombeck, J. Hébert, A. Richel, Enzymatic hydrolysis to produce nanocellulose in an integrated forest biorefinery strategy. A review, *Biotechnol. Agron. Soc. Environ.* 20 (2016) 94–103.

81. Y. Mazlita, H. V Lee, S.B.A. Hamid, Preparation of cellulose nanocrystals bio-polymer from Agro-industrial wastes: Separation and characterization, *Polym. Polym. Compos.* 24 (2016) 719–728.

82. Y. Zhang, T. Nypelö, C. Salas, J. Arboleda, I.C. Hoeger, O.J. Rojas, Cellulose nanofibrils: From strong materials to bioactive surfaces, *J. Renew. Mater.* 1 (2013). doi:10.7569/JRM.2013.634115.

83. X. Zhang, M. Rolandi, Engineering strategies for chitin nanofibers, *J. Mater. Chem. B.* (2017) 1–23. doi:10.1039/C6TB03324E.

84. Y.I. Cho, H.K. No, S.P. Meyers, Physicochemical characteristics and functional properties of various commercial chitin and chitosan products, *Journal of Agricultural and Food Chemistry* 46 (1998) 3839–3843. doi: 10.1021/jf971047f.

85. A. Percot, C. Viton, A. Domard, Optimization of chitin extraction for shrimp shells, *Biomacromolecules* 4 (2003) 12–18.

86. J. Li, J. Revol, R.H. Marchessault, Effect of degree of deacetylation of chitin on the properties of chitin crystallites, *J. Appl. Polym. Sci.* 65 (1997) 373–380.

87. J.F. Revol, R.H. Marchessault, In vitro chiral nematic ordering of chitin crystallites, *Int. J. Biol. Macromol.* 15 (1993) 329–335.

88. J. Wu, K. Zhang, N. Girouard, J.C. Meredith, Facile route to produce chitin nano fibers as precursors for flexible and transparent gas barrier materials, *Biomacromolecules* 15 (2014) 4614–4620.

89. Y. Fan, T. Saito, A. Isogai, Preparation of chitin nanofibers from squid pen β-chitin by simple mechanical treatment under acid conditions, *Biomacromolecules* 9 (2008) 1919–1923.

90. D.X. Oh, Y. Cha, H.-L. Nguyen, Y. Je, Y. Jho, D. Hwang, et al., Chiral nematic self-assembly of minimally surface damaged chitin nanofibrils and its load bearing functions, *Sci. Rep.* 6 (2016) 23245.

91. A. Dufresne, Polysaccharide nanocrystal reinforced, *Can. J. Chem.* 494 (2008) 484–494. doi:10.1139/V07-152.

92. H.-Y. Kim, D.J. Park, J.-Y. Kim, S.-T. Lim, Preparation of crystalline starch nanoparticles using cold acid hydrolysis and ultrasonication, *Carbohydr. Polym.* 98 (2013) 295–301. doi:10.1016/j.carbpol.2013.05.085.

93. H.-Y. Kim, S.S. Park, S.-T. Lim, Preparation, characterization and utilization of starch nanoparticles, *Colloids Surf. B Biointerfaces* 126 (2015) 607–620. doi:10.1016/j.colsurfb.2014.11.011.

94. H.Y. Kim, J.H. Lee, J.Y. Kim, W.J. Lim, S.T. Lim, Characterization of nanoparticles prepared by acid hydrolysis of various starches, *Starch/Staerke* 64 (2012) 367–373. doi:10.1002/star.201100105.

95. H. Angellier, L. Choisnard, S. Molina-Boisseau, P. Ozil, A. Dufresne, Optimization of the preparation of aqueous suspensions of waxy maize starch nanocrystals using a response surface methodology, *Biomacromolecules* 5 (2004) 1545–1551. doi:10.1021/bm049914u.

96. D. Song, Y.S. Thio, Y. Deng, Starch nanoparticle formation via reactive extrusion and related mechanism study, *Carbohydr. Polym.* 85 (2011) 208–214. doi:10.1016/j.carbpol.2011.02.016.

97. D. Liu, Q. Wu, H. Chen, P.R. Chang, Transitional properties of starch colloid with particle size reduction from micro- to nanometer, *J. Colloid Interface Sci.* 339 (2009) 117–124. doi:10.1016/j.jcis.2009.07.035.

98. W. Xie, Y. Zhang, Y. Liu, Homogenous carboxymethylation of starch using 1-butyl-3-methylimidazolium chloride ionic liquid medium as a solvent, *Carbohydr. Polym.* 85 (2011) 792–797. doi:10.1016/j.carbpol.2011.03.047.

99. K. Wilpiszewska, T. Spychaj, Ionic liquids: Media for starch dissolution, plasticization and modification, *Carbohydr. Polym.* 86 (2011) 424–428. doi:10.1016/j.carbpol.2011.06.001.

100. G. Zhou, Z. Luo, X. Fu, Preparation and characterization of starch nanoparticles in ionic liquid-in-oil microemulsions system, *Ind. Crops Prod.* 52 (2014) 105–110. doi:10.1016/j.indcrop.2013.10.019.

101. A. Hebeish, M.H. El-Rafie, M.A. EL-Sheikh, M.E. El-Naggar, Ultra-fine characteristics of starch nanoparticles prepared using native starch with and without surfactant, *J. Inorg. Organomet. Polym. Mater.* 24 (2014) 515–524. doi:10.1007/s10904-013-0004-x.

102. S. Bel Haaj, A. Magnin, C. Pétrier, S. Boufi, Starch nanoparticles formation via high power ultrasonication, *Carbohydr. Polym.* 92 (2013) 1625–32. doi:10.1016/j.carbpol.2012.11.022.

103. Wang, N., Ding, E., Cheng, R. Preparation and liquid crystalline properties of spherical cellulose nanocrystals, *Langmuir* 24 (2008), 5–8.

104. K. Tashiro, M. Kobayashi, Theoretical evaluation of three-dimensional elastic constants of native and regenerated celluloses: role of hydrogen bonds, *Polymer (Guildf)* 32 (1991) 1516–1526.

105. L. Petersson, K. Oksman, Biopolymer based nanocomposites: Comparing layered silicates and microcrystalline cellulose as nanoreinforcement, *Compos. Sci. Technol.* 66 (2006) 2187–2196.

106. J.D. Goodrich, W.T. Winter, Green composites prepared from cellulose nanoparticles, *ACS Symp. Ser.* 1017 (2009) 153–168.

107. Paillet, M., Dufresne, A. Chitin whisker reinforced thermoplastic nanocomposites. *Macromolecules* 34 (2001), 6527.

108. H. Sehaqui, N.E. Mushi, S. Morimune, M. Salajkova, T. Nishino, L.A. Berglund, Cellulose nanofiber orientation in nanopaper and nanocomposites by cold drawing, *ACS Appl. Mater. Interfaces* 4 (2012) 1043–1049.

109. M. Mincea, A. Negrulescu, V. Ostafe, Preparation, modification, and application of chitin nanowhiskers: a review, *Rev. Adv. Mater. Sci.* 30 (2012) 225–242.

110. H. Angellier-Coussy, J.L. Putaux, S. Molina-Boisseau, A. Dufresne, E. Bertoft, S. Perez, The molecular structure of waxy maize starch nanocrystals, *Carbohydr. Res.* 344 (2009) 1558–1566. doi:10.1016/j.carres.2009.04.002.

111. H. Angellier, S. Molina-Boisseau, M.N. Belgacem, A. Dufresne, Surface chemical modification of waxy maize starch nanocrystals, *Langmuir* 21 (2005) 2425–2433. doi:10.1021/la047530j.

112. H. Sashiwa, S.I. Aiba, Chemically modified chitin and chitosan as biomaterials, *Prog. Polym. Sci.* 29 (2004) 887–908. doi:10.1016/j.progpolymsci.2004.04.001.

113. D. Klemm, F. Kramer, S. Moritz, T. Lindström, M. Ankerfors, D. Gray, et al., Nanocelluloses: A new family of nature-based materials, 50 (2011) 5438–5466. doi:10.1002/anie.201001273.

114. Roadmap for the Development of International Standards for Nanocellulose (2011). Available at: https://www.tappinano.org/media/1468/2011-roadmapfornanocellulosestandards.pdf.

115. N. Lin, A. Dufresne, Surface modification of polysaccharide nanocrystals, polysaccharide-based nanocrystals, *Chem. Appl.* (2015) 63–108. doi:10.1002/9783527689378.ch3.

116. C. Moreau, A. Villares, I. Capron, B. Cathala, Tuning supramolecular interactions of cellulose nanocrystals to design innovative functional materials, *Ind. Crops Prod.* 93 (2015) 96–107. doi:10.1016/j.indcrop.2016.02.028.

117. Q. Zhao, S. Wang, X. Cheng, R.C.M. Yam, D. Kong, R.K.Y. Li, Surface modification of cellulose fiber via supramolecular assembly of biodegradable polyesters by the aid of host-guest inclusion complexation, *Biomacromolecules* 11 (2010) 1364–1369.

118. M. Tatsumi, Y. Teramoto, Y. Nishio, Polymer composites reinforced by locking-in a liquid-crystalline assembly of cellulose nanocrystallites, *Biomacromolecules* 13 (2012) 1584–1591. doi:10.1021/bm300310f.

119. R. Bardet, N. Belgacem, J. Bras, Flexibility and color monitoring of cellulose nanocrystal iridescent solid films using anionic or neutral polymers, *ACS Appl. Mater. Interfaces* 7 (2015) 4010–4018. doi:10.1021/am506786t.

120. M. Giese, L.K. Blusch, M.K. Khan, W.Y. Hamad, M.J. Maclachlan, Responsive mesoporous photonic cellulose films by supramolecular cotemplating, *Angew. Chemie Int. Ed.* 53 (2014) 8880–8884. doi:10.1002/anie.201402214.

121. B. Zhu, R. Merindol, A.J. Benitez, B. Wang, A. Walther, Supramolecular engineering of hierarchically self-assembled, bioinspired, cholesteric nanocomposites formed by cellulose nanocrystals and polymers, *ACS Appl. Mater. Interfaces* 8 (2016) 11031–11040. doi:10.1021/acsami.6b00410.

122. J.M. Malho, S. Arola, P. Laaksonen, G.R. Szilvay, O. Ikkala, M.B. Linder, Modular architecture of protein binding units for designing properties of cellulose nanomaterials, *Angew. Chemie Int. Ed.* 54 (2015) 12025–12028. doi:10.1002/anie.201505980.

123. J.M. Malho, C. Ouellet-Plamondon, M. Rüggeberg, P. Laaksonen, O. Ikkala, I. Burgert, et al., Enhanced plastic deformations of nanofibrillated cellulose film by adsorbed moisture and protein-mediated interactions, *Biomacromolecules* 16 (2015) 311–318. doi:10.1021/bm501514w.

124. M. V Biyani, E.J. Foster, C. Weder, Light-healable supramolecular nanocomposites based on modified cellulose nanocrystals, *ACS Macro Lett.* 2 (2013) 236–240.

125. X. Mu, D.G. Gray, Formation of chiral nematic films from cellulose nanocrystal suspensions is a two-stage process, *Langmuir* 30 (2014) 9256–9260. doi:10.1021/la501741r.

126. J.R. McKee, J. Huokuna, L. Martikainen, M. Karesoja, A. Nykänen, E. Kontturi, et al., Molecular engineering of fracture energy dissipating sacrificial bonds into cellulose nanocrystal nanocomposites, *Angew. Chemie Int. Ed.* 53 (2014) 5049–5053. doi:10.1002/anie.201401072.

127. J.M. Raquez, A.-L. Goffin, E. Duquesne, Y. Habibi, A. Dufresne, P. Dubois, Novel nanocomposites reinforced with polysaccharide (Starch) nanocrystals: from interfacial ring-opening polymerization to melt-processing implementation, in: P.B. Smith, R.A. Gross (Eds.), *Biobased Monomers, Polymers, and Materials,* ACS Sympos, American Chemical Society, Washington, D.C., 2012: pp. 257–268. doi:10.102/bk-2012-1105.ch016.

128. J. Yu, F. Ai, A. Dufresne, S. Gao, J. Huang, P.R. Chang, Structure and mechanical properties of poly(lactic acid) filled with (starch nanocrystal)-graft-poly(ε-caprolactone), *Macromol. Mater. Eng.* 293 (2008) 763–770. doi:10.1002/mame.200800134.

129. H. Namazi, A. Dadkhah, Surface modification of starch nanocrystals through ring-opening polymerization of ε-caprolactone and investigation of their microstructures, *J. Appl. Polym. Sci.* 110 (2008) 2405–2412. doi:10.1002/app.

130. W. Thielemans, M.N. Belgacem, A. Dufresne, Starch nanocrystals with large chain surface modifications, *Langmuir* 22 (2006) 4804–4810. doi:10.1017/CBO9781107415324.004.

131. Y. Xu, W. Ding, J. Liu, Y. Li, J.F. Kennedy, Q. Gu, et al., Preparation and characterization of organic-soluble acetylated starch nanocrystals, *Carbohydr. Polym.* 80 (2010) 1078–1084. doi:10.1016/j.carbpol.2010.01.027.

132. P.R. Chang, F. Ai, Y. Chen, A. Dufresne, J. Huang, Effects of starch nanocrystal-graft-porycaprolactone on mechanical properties of waterborne polyurethane-based nanocomposites, *J. Appl. Polym. Sci.* 111 (2009) 619–627. doi:10.1002/app.29060.

133. Z. Yang, H. Peng, W. Wang, T. Liu, Crystallization behavior of poly(ε-caprolactone)/layered double hydroxide nanocomposites, *J. Appl. Polym. Sci.* 116 (2010) 2658–2667. doi:10.1002/app.

134. M. Labet, W. Thielemans, A. Dufresne, Polymer grafting onto starch nanocrystals, *Biomacromolecules* 8 (2007) 2916–2927. doi:10.1021/bm700468f.

135. H. Namazi, F. Fathi, A. Dadkhah, Hydrophobically modified starch using long-chain fatty acids for preparation of nanosized starch particles, *Sci. Iran.* 18 (2011) 439–445. doi:10.1016/j.scient.2011.05.006.

136. Y. Chen, F. Liu, X. Deng, D. Jia, Y. Luo, Structure and properties of comb-branched waterborne polyurethane/montmorillonite nanocomposites, *Huagong Xuebao/Journal of Chemical Industry and Engineering* (Chinese Ed. 59 (2008) 761–765.

137. S. Benali, F. Khelifa, D. Lerari, R. Mincheva, Y. Habibi, D. Lahem, et al., Supramolecular approach for an efficient processing of polylactide/starch nanocomposites, *ACS Omega* 3 (2018), 1069–1080.

138. A. Ilnicka, J.P. Lukaszewicz, Discussion remarks on the role of wood and chitin constituents during carbonization, *Front. Mater.* 2 (2015) 1–6. doi:10.3389/fmats.2015.00020.

139. J. Li, J. Revol, R. Marchessault, Effect of degree of deacetylation of chitin on the properties of chitin crystallites, J. Appl. Polym. (1997) 373–380. doi:10.1002/(SICI)1097-4628(19970711)65:2<373::AID-APP18>3.0.CO;2-0.

140. K.G. Nair, A. Dufresne, A. Gandini, M.N. Belgacem, Crab shell chitin whiskers reinforced natural rubber nanocomposites. 3. Effect of Chemical Modification of chitin whiskers, *Biomacromolecules* 4 (2003) 1835–1842. doi:10.1021/bm030058g.

141. C. Guo, L. Zhou, J. Lv, Effects of expandable graphite and modified ammonium polyphosphate on the flame-retardant and mechanical properties of wood flour-polypropylene composites, *Polym. Polym. Compos.* 21 (2013) 449–456. doi:10.1002/app.

142. A. Dufresne, Processing of polymer nanocomposites reinforced with polysaccharide nanocrystals, *Molecules* 15 (2010) 4111–4128. doi:10.3390/molecules15064111.

143. T. Setoguchi, K. Yamamoto, J.I. Kadokawa, Preparation of chitin nanofiber-graft-poly(l-lactide-co-ε-caprolactone) films by surface-initiated ring-opening graft copolymerization, *Polymer* (United Kingdom) 53 (2012) 4977–4982. doi:10.1016/j.polymer.2012.08.060.

144. S. Ifuku, M. Iwasaki, M. Morimoto, H. Saimoto, Graft polymerization of acrylic acid onto chitin nanofiber to improve dispersibility in basic water, *Carbohydr. Polym.* 90 (2012) 623–627. doi:10.1016/j.carbpol.2012.05.087.

145. H. Liu, W. Liu, B. Luo, W. Wen, M. Liu, X. Wang, et al., Electrospun composite nanofiber membrane of poly(l-lactide) and surface grafted chitin whiskers: Fabrication, mechanical properties and cytocompatibility, *Carbohydr. Polym.* 147 (2016) 216–225. doi:10.1016/j.carbpol.2016.03.096.

146. Z. Zou, C. Luo, B. Luo, W. Wen, M. Liu, C. Zhou, Synergistic reinforcing and toughening of poly(L-lactide) composites with surface-modified MgO and chitin whiskers, *Compos. Sci. Technol.* 133 (2016) 128–135. doi:10.1016/j.compscitech.2016.07.025.

147. A. De Sousa Mol, I. Martins, R.L. Oréfice, Surface-pegylated chitin whiskers as an effective additive to enhance the mechanical properties of recycled ABS, *J. Appl. Polym. Sci.* 132 (2015) 1–8. doi:10.1002/app.42463.

148. J. Wang, Z. Wang, J. Li, B. Wang, J. Liu, P. Chen, et al., Chitin nanocrystals grafted with poly(3-hydroxybutyrate-co-3-hydroxyvalerate) and their effects on thermal behavior of PHBV, *Carbohydr. Polym.* 87 (2012) 784–789. doi:10.1016/j.carbpol.2011.08.066.

Index